Studies in Computational Intelligence

Volume 722

Series editor

Janusz Kacprzyk, Polish Academy of Sciences, Warsaw, Poland
e-mail: kacprzyk@ibspan.waw.pl

About this Series

The series "Studies in Computational Intelligence" (SCI) publishes new developments and advances in the various areas of computational intelligence—quickly and with a high quality. The intent is to cover the theory, applications, and design methods of computational intelligence, as embedded in the fields of engineering, computer science, physics and life sciences, as well as the methodologies behind them. The series contains monographs, lecture notes and edited volumes in computational intelligence spanning the areas of neural networks, connectionist systems, genetic algorithms, evolutionary computation, artificial intelligence, cellular automata, self-organizing systems, soft computing, fuzzy systems, and hybrid intelligent systems. Of particular value to both the contributors and the readership are the short publication timeframe and the worldwide distribution, which enable both wide and rapid dissemination of research output.

More information about this series at http://www.springer.com/series/7092

Roger Lee

Editor

Software Engineering Research, Management and Applications

 Springer

Editor
Roger Lee
Software Engineering and Information
  Technology Institute
Central Michigan University
Mount Pleasant, MI
USA

ISSN 1860-949X   ISSN 1860-9503   (electronic)
Studies in Computational Intelligence
ISBN 978-3-319-87069-4   ISBN 978-3-319-61388-8   (eBook)
DOI 10.1007/978-3-319-61388-8

Printed on acid-free paper

This Springer imprint is published by Springer Nature
The registered company is Springer International Publishing AG
The registered company address is: Gewerbestrasse 11, 6330 Cham, Switzerland

Foreword

The purpose of the 15th International Conference on Software Engineering, Artificial Intelligence Research, Management and Applications (SERA 2017) held on 7–9 June 2016 at the University of Greenwich, UK, is aimed at bringing together scientists, engineers, computer users, and students to share their experiences and exchange new ideas and research results about all aspects (theory, applications, and tools) of Software Engineering Research, Management, and Applications, and to discuss the practical challenges encountered along the way and the solutions adopted to solve them. The conference organizers selected the best 12 papers from those papers accepted for presentation at the conference in order to publish them in this volume. The papers were chosen based on review scored submitted by members of the program committee and underwent further rigorous rounds of review.

In Chap. "Agile Web Development Methodologies: A Survey and Evaluation", Nasrin Ghasempour Maleki and Raman Ramsin provide a criteria-based evaluation of fourteen agile Web development methodologies. The evaluation results highlight the strengths and weaknesses of the methodologies as to their general processes, modelling languages, agile features, and Web development facilities and can, therefore, help Web developers choose the methodology that best fits their project needs.

In Chap. "Load Experiment of the vDACS Scheme in Case of Increasing the Simultaneous Connection for the DACS SV", Kazuya Odagiri, Shogo Shimizu, and Naohiro Ishii perform a load experiment of the cloud type virtual PBNM named the vDACS Scheme, which can be used by plural organizations, for applications to the small- and medium-size scale organizations.

In Chap. "Blind Channel Estimation Using Novel Independent Component Analysis with Pulse Shaping for Interference Cancellation", Renuka Bhandari and Sangeeta Jadhav designing the novel blind channel estimation approach using independent component analysis (ICA) with both ISI cancellation and blind interference cancellation. This method is named as hybrid ICA (HICA).

In Chap. "Anticipated Test Design and its Application to Evaluate and Select Embedded Libraries", Clauirton Siebra, Carla Nascimento, Leonardo Sodre, Antônio Cavalcanti, Daniel Barros, Fernando Lima, Fernando Cruz, Fábio Q. B. da Silva, and Andre L M Santos present an anticipated test design methodology; their work applies this strategy to the development of a set of libraries that are used in several other projects.

In Chap. "Improving Web Application Reliability and Testing Using Accurate Usage Models", Gity Karami and Jeff Tian examine the impact of accurate usage models on reliability, test coverage, and test efficiency. A case study is carried out to quantify this impact. They found supporting evidence that accurate Markov OP improves reliability, test coverage, and test efficiency.

In Chap. "C-PLAD-SM: Extending Component Requirements with Use Cases and State Machines", Kevin A. Gary and M. Brian Blake describe an extension to the C-PLAD approach, dubbed C-PLAD-SM, which addresses the gaps in their earlier work.

In Chap. "A Structural Rule-Based Approach for Design Patterns Recovery", Mohammed Ghazi Al-Obeidallah, Miltos Petridis, and Stelios Kapetanakis present a multiple levels detection approach (MLDA) to recover design pattern instances from Java source code. MLDA is able to extract design pattern instances based on a generated class-level representation of an investigated system.

In Chap. "DRSS: Distributed RDF SPARQL Streaming", Amadou Fall Dia, Zakia Kazi-Aoul, Aliou Boly, and Elisabeth Metais present DRSS, a distributed and scalable engine for RDF streams processing. DRSS proposes a new query syntax for continuous querying of RDF data streams.

In Chap. "An Efficient Approach for Real-Time Processing of RDSZ-Based Compressed RDF Streams", Ndeye Bousso Deme, Amadou Fall Dia, Aliou Boly, Zakia Kazi-Aoul, and Raja Chiky propose an approach for continuous querying RDSZ-based RDF streams without decompression phase. They add three algorithms from simple to aggregate query execution over RDSZ-compressed items.

In Chap. "Energy Efficiency Cluster Head Election Using Fuzzy Logic Method for Wireless Sensor Networks", Wided Abidi and Tahar Ezzedine introduce a new clustering algorithm which elects CHs using fuzzy logic method and based on a set of parameters which increases the lifetime of WSN.

In Chap. "Enabling GSD Task Allocation via Cloud-Based Software Processes", Sami Alajrami, Barbara Gallina, and Alexander Romanovsky propose to integrate and semi-automate the calculation of an existing global distance metric (GDM) into an architecture that supports executing cloud-based software processes.

In Chap. "Composite Event Handling over a Distributed Event-Based System", Amina Chaabane, Salma Bradai, Wassef Louati, and Mohamed Jmaiel address the structured peer-to-peer network shortcomings. They exploit advantages offered by structured topology (distributed hash table DHT) and extend it by novel approach in order to improve expressiveness by supporting complex event processing (CEP).

It is our sincere hope that this volume provides stimulation and inspiration, and that it will be used as a foundation for works to come.

June 2017

Program Chairs:
Lachlan MacKinnon
Jixin Ma
University of Greenwich, London, UK

Contents

Contributors

Wided Abidi Engineering School of Tunis, Communications Systems Laboratory, University of Tunis El Manar, Tunis, Tunisia

Mohammed Ghazi Al-Obeidallah Department of Computing, University of Brighton, Brighton, UK

Sami Alajrami Newcastle University, Newcastle upon Tyne, UK

Renuka Bhandari Department of E&TC, Dr. D.Y. Patil Institute of Engineering & Technology, Pune, India; Army Institute of Technology Pune, Pune, India

Daniel Barros CIn/Samsung Laboratory of Research and Development, Recife, Brazil

M.B. Blake College of Computing & Informatics, Drexel University, Philadelphia, PA, USA

Aliou Boly LID Lab, UCAD, Dakar-Fann, Senegal

Salma Bradai ReDCAD Laboratory, University of Sfax, National School of Engineers of Sfax, Sfax, Tunisia

Antônio Cavalcanti CIn/Samsung Laboratory of Research and Development, Recife, Brazil

Amina Chaabane Higher Institute of Applied Sciences and Technology, University of Kairouane, Kasserine, Tunisia

Raja Chiky LISITE Lab, ISEP, Paris, France

Fernando Cruz CIn/Samsung Laboratory of Research and Development, Recife, Brazil

Amadou Fall Dia LISITE Lab, ISEP, Paris, France

Ndéye Bousso Déme LID Lab, UCAD, Dakar-Fann, Senegal

Tahar Ezzedine Engineering School of Tunis, Communications Systems Laboratory, University of Tunis El Manar, Tunis, Tunisia

Barbara Gallina MÃ¤laradalen Univeristy, VÃ¤sterÃ¥s, Sweden

Kevin A. Gary The School of Computing Informatics, and Decision Systems Engineering, The Ira A. Fulton Schools of Engineering, Arizona State University, Mesa, AZ, USA

Naohiro Ishii Aichi Institute of Technology, Toyota, Aichi, Japan

Sangeeta Jadhav Army Institute of Technology Pune, Pune, India

Mohamed Jmaiel Research Center for Computer Science, Multimedia and Digital Data Processing of Sfax, Sfax, Tunisia

Stelios Kapetanakis Department of Computing, University of Brighton, Brighton, UK

Gity Karami Department of Computer Science and Engineering, Southern Methodist University, Dallas, TX, USA

Zakia Kazi-Aoul LISITE Lab, ISEP, Paris, France

Fernando Lima CIn/Samsung Laboratory of Research and Development, Recife, Brazil

Wassef Louati Faculty of Economics and Management of Sfax, University of Sfax, Sfax, Tunisia

Nasrin Ghasempour Maleki Department of Computer Engineering, Sharif University of Technology, Tehran, Iran

Elisabeth Métais CEDRIC Lab, CNAM, Paris, France

Carla Nascimento CIn/Samsung Laboratory of Research and Development, Recife, Brazil

Kazuya Odagiri Sugiyama Jogakuen University, Nagoya, Aichi, Japan

Miltos Petridis Department of Computing, Middlesex University, London, UK

Raman Ramsin Department of Computer Engineering, Sharif University of Technology, Tehran, Iran

Alexander Romanovsky Newcastle University, Newcastle upon Tyne, UK

Andre L.M. Santos Centro de Informática, Universidade Federal de Pernambuco, Recife, Brazil

Shogo Shimizu Gakushuin Women's College, Tokyo, Japan

Clauirton Siebra Informatics Center, Federal University of Paraiba, Joao Pessoa, Brazil

Fábio Q.B. da Silva Centro de Informática, Universidade Federal de Pernambuco, Recife, Brazil

Leonardo Sodre CIn/Samsung Laboratory of Research and Development, Recife, Brazil

Jeff Tian Department of Computer Science and Engineering, Southern Methodist University, Dallas, TX, USA; School of Computer Science, Northwestern Polytechnical University, Xi'an, Shaanxi, China

Agile Web Development Methodologies: A Survey and Evaluation

Nasrin Ghasempour Maleki and Raman Ramsin

Abstract Dynamic and accessible web systems have gained utmost importance in modern life. Due to the competitive nature of such systems, they need to be superior as to performance, scalability, and security. Web systems typically require short time-to-markets, and it should be possible to easily implement new requirements into working web systems. These ideals have made agile methods especially suitable for developing such systems, as they promote productivity, facilitate continuous interaction with customers, and enhance the flexibility and quality of the software produced. When starting a web development project, selecting the methodology that fits the project situation can be an important factor in the ultimate success of the endeavor. In order to facilitate the selection process, we provide a criteria-based evaluation of fourteen agile web development methodologies. The evaluation results highlight the strengths and weaknesses of the methodologies as to their general processes, modeling languages, agile features, and web development facilities, and can therefore help web developers choose the methodology that best fits their project needs.

Keywords Software development methodology · Agile method · Web system · Web development methodology · Criteria-based evaluation

1 Introduction

Businesses increasingly rely on web systems for maintaining their competitive edge, and the widespread use of these systems has made them indispensable in everyday life. Due to their pivotal role, web systems have to be developed fast, and

N.G. Maleki · R. Ramsin (✉)
Department of Computer Engineering,
Sharif University of Technology, Tehran, Iran
e-mail: ramsin@sharif.edu

N.G. Maleki
e-mail: ghasempourmk@alum.sharif.edu

they should be flexible enough to be easily changed and extended as required; also, special attention should be given to proper requirements engineering and continuous verification/validation of these systems. An important feature of web development projects is their highly dynamic nature, which necessitates constant user feedback. Due to the above characteristics, web development involves much more than mere web "programming": developers have thus realized that using the right software development methodology is essential for successful construction and evolution of web systems.

Agile methodologies are suitable candidates for developing web systems, since they adequately address the specific needs of this context. However, there are many agile web development methodologies to choose from, and choosing the right one can be a serious challenge for web development teams. Making the right choice requires adequate knowledge about the strengths and weaknesses of each methodology; however, development teams should not be expected to acquire this knowledge through hands-on experience with each and every methodology. Fortunately, criteria-based evaluation of methodologies is a proven method for identifying and accentuating the capabilities and limitations of software development methodologies. Several such evaluations have previously been conducted on various types of methodologies [1–3], but the need remains for a comprehensive evaluation of modern agile web development methodologies.

We provide a comprehensive criteria-based evaluation of fourteen prominent agile web development methodologies. Methodologies have been targeted for evaluation based on their popularity and documentation; methodologies that lack proper methodology documentation (on the process, products, and people involved) have not been included. The evaluation criteria have been collected from multiple sources, and have been adapted to the specific characteristics of the agile web development context. Evaluation results clearly show the pros and cons of the methodologies, and can be used by web developers to choose the methodology that fits their needs.

The rest of the paper is structured as follows: Sect. 2 provides a brief overview of the targeted agile web development methodologies; Sect. 3 presents the evaluations criteria; Sect. 4 lists the results of applying the evaluation criteria to the methodologies; and Sect. 5 presents the conclusions and suggests ways for furthering this research.

2 An Overview of Targeted Methodologies

The fourteen agile web development methodologies targeted for evaluation have been briefly introduced throughout the rest of this section.

2.1 MockupDD

MockupDD is an agile model-driven web engineering methodology based on Scrum [4]; its process consists of four phases (Fig. 1):

1. *Mockup Construction*: Requirements are gathered from the collection of stories by customers or final users through using mockups to produce graphical stories.
2. *Mockup Processing*: Important parts of the UI are identified through mapping the basic concepts of mockups to a structural UI meta-model.
3. *Features specification and tags refinement*: Mockups are tagged with labels that represent their semantics. User stories are then adapted with the mockups, and the tags are classified.
4. *Code and Model Generation*: Tags will either be converted into web engineering elements, or be combined to identify more complex design features. After the full definition of tags, an executable version is produced; other models of model-driven web engineering are created based on this version.

2.2 RAMBUS

RAMBUS is an agile methodology loosely based on Scrum [5]; its process consists of three phases (Fig. 2):

1. *Communication*: Communication with users is performed to capture the functional requirements on story cards. To show the behavior of the system, a

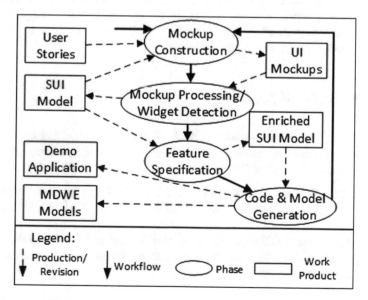

Fig. 1 Process of MockupDD

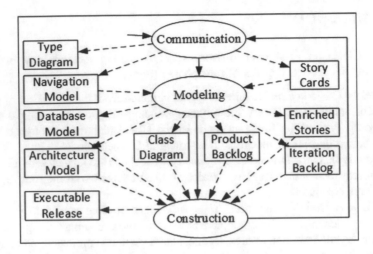

Fig. 2 Process of RAMBUS

navigation model is created. For each story card, user priorities, predicted difficulties in implementing the story, and the relevant items of the navigation model are written on the back of the card. Type diagrams are produced to show the relationships of the elements.

2. *Modeling*: Class and type diagrams are developed/refined iteratively, and a database model is created. User stories are enriched with user acceptance criteria. Reuse options are explored, and nonfunctional requirements are considered in the user stories.

3. *Construction*: Coding and testing are performed, resulting in an executable release. Daily sessions, strict coding standards, test-driven development, continuous integration, and pair programming are the agile practices prescribed by the methodology for this particular phase.

2.3   USABAGILE_Web

USABAGILE_Web is a methodology for designing or reengineering a web system by architectural analysis, creating a UI prototype, and usability testing [6]. Before the main process, three usability assessment activities are performed:

1. *Inspection*: UI structure is inspected to detect usability problems. Typically, a team of 3–5 specialists performs Nielsen analysis. UI functionality is not considered.

2. *Evaluation*: Under the supervision of experts, the usability of web pages is analyzed based on components such as links, forms, and the elements with which the user interacts.

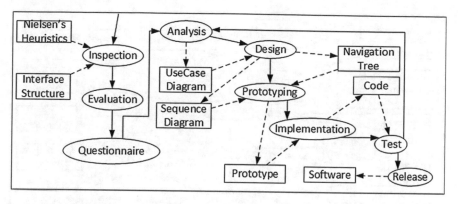

Fig. 3 Process of USABAGILE_Web

3. *Questionnaire*: A questionnaire is used for capturing the wishes and feelings of users after using the UI.

The results of the above activities are documented in a special usability report. The main process uses this usability report as input, and consists of six phases (Fig. 3):

1. *Analysis*: UI behavior is captured in behavioral use case diagrams.
2. *Design*: A summary of the UI structure related to user operations is produced for logical analysis. Navigation features of the UI are shown to the customer, and the feedback is used for analyzing the UI design.
3. *Prototyping*: This phase is integrated with the two previous phases. UI prototypes are created by experts based on analysis and design results.
4. *Implementation*: After UI prototypes are accepted by the customers and experts, the system is implemented.
5. *Test*: A set of potential users are selected (preferably from among those who filled the assessment questionnaire) to test the new UI. New features are implemented as required, and the process is iterated until the product is fully validated by the users.
6. *Release*: The produced/reengineered web system is deployed into the user environment.

2.4 Augmented WebHelix

WebHelix was introduced in 2006 as a spiral lightweight methodology for teaching web development to students [7]. Augmented WebHelix is a practical, business-oriented web development methodology that extends WebHelix with management and Q/A activities [8]; its main process consists of eight phases (Fig. 4):

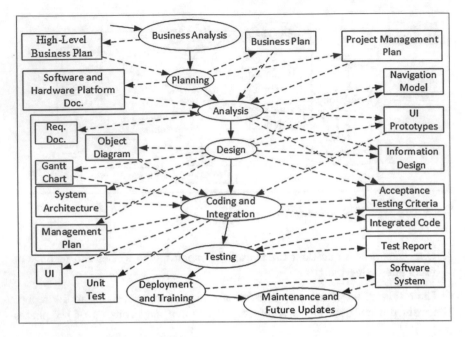

Fig. 4 Process of Augmented WebHelix

1. *Business Analysis*: Spans identifying business processes, identifying real and virtual chains of supply, and providing a high-level business plan.
2. *Planning*: Spans identifying the software and hardware platforms, specifying the project management scheme and the necessary tools and resources, and producing a business plan.
3. *Analysis*: Spans creating or updating the requirements, creating/updating the navigation model, creating UI prototypes, creating/updating the information structure, and identifying criteria for acceptance testing.
4. *Design*: Spans creating or updating a detailed system architecture, updating the system UI and navigation model, creating a system object diagram, creating or updating the system information design, creating or updating the management plan, forming the programming team, creating a Gantt chart, and identifying test criteria for system acceptance.
5. *Coding and Integration*: Spans components selection, implementing the UI, coding, integration, unit testing, code review, and updating the acceptance criteria.
6. *Testing*: Spans web design testing, multimedia testing, and user acceptance testing.
7. *Deployment and Training*: The system is deployed into the network environment, and the users are trained.
8. *Maintenance and Future Updates*: Spans maintaining and updating the system.

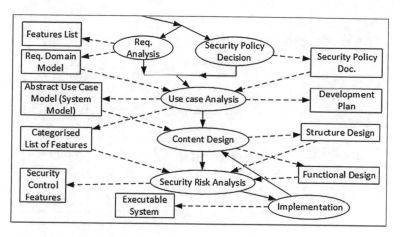

Fig. 5 Process of Secure FDD

2.5   Secure FDD

Secure FDD extends the Feature-Driven Development (FDD) methodology with security analysis and design features in order to develop secure web systems [9]; its process consists of six stages (Fig. 5):

1. *Requirements Analysis*: Security-related needs and expectations of the stake-holders are identified, and security rules are set. A list of features (as defined in FDD) is also produced.
2. *Security Policy Decision*: Policies on how to implement security are specified. These policies help build the web system in a security-conscious manner.
3. *Use Case Analysis*: Features are classified and an overall structural model is produced. Use case analysis is performed for refining the system scope.
4. *Content Design*: A blueprint for implementing the features is produced by conducting structural design (focusing on feature content) and functional design (focusing on the user actions involved in each feature).
5. *Security Risk Analysis*: An iterative-incremental process is performed to determine security control features.
6. *Implementation*: The target system is implemented, with special attention to security features.

2.6   XWebProcess

XWebProcess extends the Extreme Programming (XP) methodology with web development features [10]; its process consists of six stages (Fig. 6):

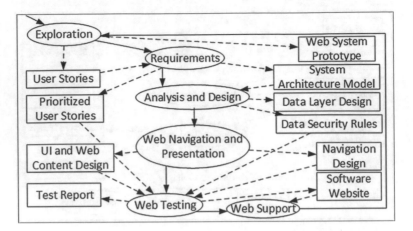

Fig. 6 Process of XWebProcess

1. *Exploration*: High-level requirements are captured in user stories, and the overall system design is determined by prototyping.
2. *Requirements*: The system architecture is defined, with special attention to flexibility, efficiency, and maintainability. User stories are estimated and prioritized, and high-priority stories are selected for development in the next cycle.
3. *Analysis and Design*: The data layer is designed based on the data recovery and security rules added to XP.
4. *Web Navigation and Presentation*: Web content and navigation is designed by using the design practices added to XP. The web system is then implemented.
5. *Web Testing*: Verification and validation are performed, and detected bugs are fixed. Support analysts assist the developers if a special setup configuration (e.g., files, devices, and environment variables) is required for running the tests.
6. *Web Support*: Website components are maintained.

2.7 XP

The XP methodology (in its original, non-extended form) can also be effectively used for developing web systems [11, 12]; its process consists of six phases (Fig. 7):

1. *Exploration*: Activities include team formation, elicitation of high-level requirements (as user stories), and specification of system architecture.
2. *Planning*: User stories are estimated, prioritized, and broken down into development tasks for programmers to complete in 1–3 weeks. A subset of the stories is then selected for implementation in the first release.

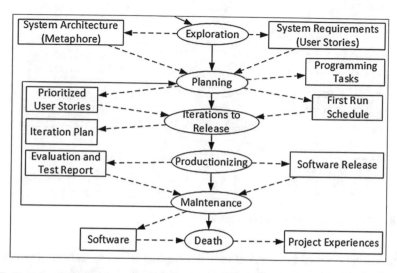

Fig. 7 Process of XP

3. *Iterations to Release*: Analysis, design, coding, testing and integration are performed iteratively in a collective code ownership environment.
4. *Productionizing*: System-wide testing is performed, and the system is deployed into the user environment.
5. *Maintenance*: The remaining user stories are implemented by repeating phases 2, 3 and 4.
6. *Death*: Project review and post-mortem are conducted.

2.8 UML-Based Agile Method

The agile web development method proposed by Lee et al. involves modeling activities using an extension of UML [13]. The produced UML model for the web application consists of a navigation model, a components communication model, a conceptual model, and an architectural model. Developers can thus take advantage of both model-based and test-based development. Its cyclic process consists of two phases (Fig. 8):

1. *Analysis*: Requirements analysis is performed, and the conceptual model is produced (as a class diagram). An architecture is also defined (as a component diagram).
2. *Construction*: This phase consists of two sub-phases: Build and Sophistication. During Build, developers iteratively select a subset of the requirements and

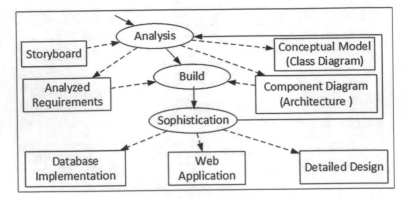

Fig. 8 Process of UML-Based Agile Method

build a storyboard that depicts how the requirements are realized through user interactions. Sophistication involves detailed design and implementation.

The implemented components are integrated with existing subsystems, and integration/regression tests are applied. The cycle is repeated until all the requirements are satisfied.

2.9 Crystal Orange Web

Crystal Orange Web is a variant of Cockburn's Crystal Orange methodology specifically designed for ongoing web development projects [14]. It stresses the importance of collaboration among the developers, and makes extensive use of agile practices. Instead of providing a specific lifecycle, the methodology prescribes five agile conventions: "Regular Heartbeat with Learning", "Basic Process", "Maximum Progress, Minimum Distractions", "Maximally Defect Free", and "A Community Aligned in Conversation"; these conventions facilitate the development and constant evolution of a web system over an extended period of time.

2.10 S-Scrum

S-Scrum is a variant of Scrum aimed at developing secure web systems [15]. The objective is to provide critical security web services and perform security analysis and design during early stages of Scrum. The methodology accommodates changing requirements; moreover, if the changed requirement is a critical security requirement, the current sprint is cut short and a new sprint is started in order to implement the changed requirement. The process of S-Scrum is analogous to the

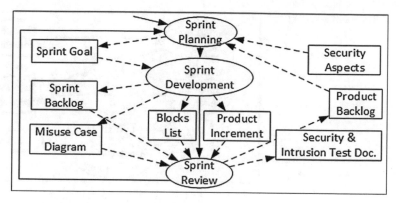

Fig. 9 Process of S-Scrum

original Scrum process; however, special attention is given to developing and applying security and intrusion tests, and producing a misuse case diagram (Fig. 9).

2.11 Scrum for CMMI Level 2

Salinas et al. have proposed an extended variant of Scrum to accommodate CMMI-Level 2 in the context of web development [16]. The methodology claims to have achieved this by adding a time-boxed "Sprint 0" at the beginning of the Scrum process (Fig. 10). "Sprint 0" deals with quality assurance, project data management, and project evaluation. After "Sprint 0", the original Scrum process is enacted along with the proposed extensions: project data is collected during Scrum meetings, and project reports are produced at the end of each sprint.

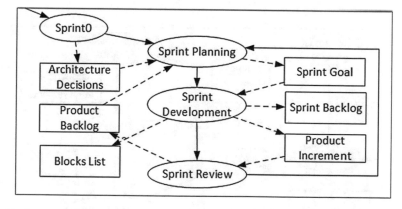

Fig. 10 Process of Scrum for CMMI Level 2

Fig. 11 Process of AWDWF

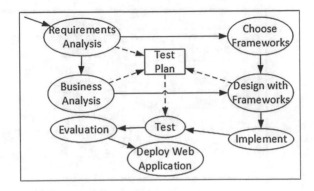

2.12 AWDWF

As the name suggests, AWDWF (Agile Web Development with Web Framework) is the result of integrating Web Framework features with the Agile Web Development process, with the specific aim of achieving fast response to requests and quick adaptation to change [17]; an analysis conducted on web development with AWDWF has shown that productivity and quality are improved. The process of the methodology consists of eight phases (Fig. 11): Requirements Analysis, Business Analysis, Choose Frameworks, Design with Frameworks, Implement, Test, Evaluation, and Deploy Web Application. The Web Framework provides a simple MVC-based programming model that can shorten the development cycle.

2.13 AWE

The iterative process of the Agile Web Engineering (AWE) methodology [18] consists of six stages (Fig. 12): Business Analysis, Requirements Analysis, Design,

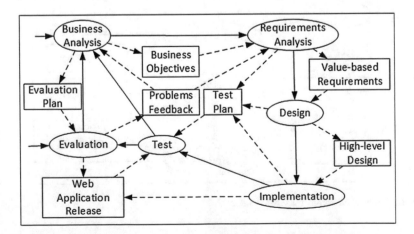

Fig. 12 Process of AWE

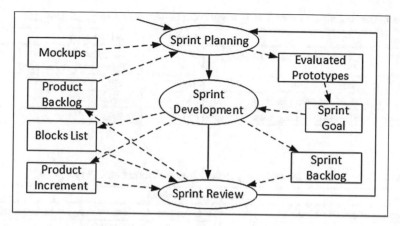

Fig. 13 Process of MDE-Scrum

Implementation, Test, and Evaluation. During the Design stage, a high-level implementation is produced that addresses all architectural issues. This version is evolved into a release of the system during Implementation and Test. Evaluation involves design-independent appraisal by the ambassador user and developers.

2.14 MDE-Scrum

This mockup-based methodology combines Model-Driven Engineering (MDE) with Scrum [19]. At the start of the process (Fig. 13), requirements are captured in user stories and mockups of the system are designed. User-approved mockups are converted into annotated UML models of the desired system. A functional prototype of the system is then generated based on these models, and is converted into an executable release. Major conversions are automated through the MockupToME tool. The Scrum process facilitates the conversion process by focusing the effort on specific high-priority features, and by supporting user-centered refinement of mockups and models.

3 Evaluation Criteria

The evaluation criteria were collected from various sources, including [1–3, 20]. They have been grouped based on the methodology feature that they evaluate. There are four groups of criteria: *modeling language, process, agility,* and *web-based features;* the criteria belonging to these categories are described in Tables 1, 2, 3 and 4, respectively.

Table 1 General criteria for evaluating methodologies—Modeling language group [1]

Name	Type	Possible values
Support for specific modeling language	SC	1: Not prescribed/enforced; 2: Prescribed; 3: Enforced
Simplicity to learn and use	SM	Yes/No
Expressiveness of modeling language	SM	Yes/No
Support for complexity management	SM	Yes/No

Table 2 General criteria for evaluating methodologies—Process group [1]

Name	Type	Possible values
Coverage of generic lifecycle	SC	D: Definition; C: Construction; M: Maintenance
Support for seamless transition between phases	SC	1: No; 2: Potentially; 3: Yes
Support for smooth transition between phases	SC	1: No; 2: Potentially; 3: Yes
Type of lifecycle	D	Waterfall (W.), Iterative-Incremental (I.-I.), etc.
Attention to design activities	SM	Yes/No
Potential of integration with other methodologies	SC	Integration strategy: 1: Not required; 2: Required but not provided; 3: Provided
Adequacy of products	SC	Relevant products in: 1: No phases; 2: Some phases; 3: All phases
Consistency of products	SC	1: Products overlap; 2: Products do not overlap
Support for modeling different views in products	SC	S: Structural; F: Functional; B: Behavioral
Support for modeling different granularity levels in products	SC	S: System; P: Package; C: Component; O: Object; D: Domain; SD: Sub-Domain; PR: Product; F: Features
Support for modeling different abstraction levels in products	SC	A: Analysis; D: Design; I: Implementation
Testability of products	SC	1: Not addressed; 2: Partial; 3: High
Tangibility of products (to customer and/or development team)	SC	1: None tangible; 2: Some not tangible to team members; 3: Some not tangible to customer; 4: All tangible
Traceability of products to requirements	SM	Yes/No
Definition of roles	SC	1: Roles not defined; 2: Roles defined, but without responsibilities; 3: Both roles and responsibilities defined
Required team knowledge/experience	SM	Yes/No
Support for team motivation mechanisms	SM	Yes/No
Expressiveness of process	SC	1: No; 2: To some extent; 3: Yes

(continued)

Table 2 (continued)

Name	Type	Possible values
Completeness of process definition	SC	L: Lifecycle; A: Activities; TP: Techniques/Practices; R: Roles; P: Products; U: Umbrella Activities; RL: Rules; ML: Modeling Language
Rationality and consistency of activities	SC	1: Problems in consistency and rationality; 2: Problems in consistency; 3: Problems in rationality; 4: No problems
Support for complexity management in process	SM	Yes/No
Attention to detail in process definition	SC	Details provided for: 1: No phases; 2: Some of the phases and internal tasks; 3: All phases and internal tasks
Definition of phase inputs and outputs (I/O)	SC	1: I/O not defined; 2: I/O defined implicitly; 3: I/O explicitly defined for all phases
Availability of documentation on process	SM	Yes/No
Tool support for process	SM	Yes/No
Ease of use of process	SC	1: Weak; 2: Average; 3: Good
Availability of experience reports of practical use	SM	Yes/No
Configurability of process	SC	1: No; 2: Possible, but not addressed explicitly; 3: Explicitly addressed
Flexibility of process	SC	1: No; 2: Possible, but not addressed explicitly; 3: Explicitly addressed
Specification of criticality level addressed by process	SC	1: Defined explicitly; 2: Not defined explicitly, but can be inferred; 3: Not defined and cannot be inferred
Platform-adaptivity of process	SM	Yes/No
Support for formalism	SM	Yes/No
Support for scalability	SC	1: Small; 2: Medium; 3: Large
Support for modularity	SM	Yes/No
Support for requirements elicitation	SC (D)	M: Uses conventional methods (description); D: Uses a specific method (description); N: No certain way
Support for requirements specification	D	
Support for requirements-based process	SM	Yes/No
Support for requirements prioritization	SM	Yes/No
Need for observation of specific constraints/assumptions	SC	1: Constraints/Assumptions exist 2: Constraints/Assumptions prescribed 3: No constraints/assumptions

Table 3 Criteria related to agility characteristics [1]

Name	Type	Possible values
Support for early and continuous delivery of working software	SC	1: Neither early nor continuous; 2: Continuous but not early; 3: Early and continuous
Support for active user involvement	SM	Yes/No
Support for continuous customer feedback	SM	Yes/No
Support for self-organizing teams	SC	1: Not discussed; 2: Addressed; 3: Ignored
Support for face-to-face conversation	SM	Yes/No
Support for velocity monitoring and control	SM	Yes/No
Attention to team behavior/efficiency	SM	Yes/No
Task assignment method	D	Voluntary sign up, Team-assigned, Manager-assigned, etc.
Support for continuous integration	SM	Yes/No
Modeling coverage	SM	Yes/No
Support for standards	SM	Yes/No
Support for iterative-incremental process	SM	Yes/No
Support for agile techniques	SM	Yes/No
Support for requirements flexibility	SM	Yes/No
Support for rapid production of artifacts	SC	1: No; 2: To some extent; 3: Yes
Support for lean development (through short time spans, and the use of tools)	SM	Yes/No
Support for learning (from previous iterations/projects)	SC	1: Not addressed; 2: Addressed implicitly; 3: Addressed explicitly
Provision of feedback by process	SM	Yes/No

The evaluation framework has been validated according to the four meta-criteria defined in [21]; validation shows that the proposed criteria are *general* enough to be applied to all agile web development methodologies, *precise* enough to help identify their similarities and differences, *comprehensive* enough to cover their important characteristics, and *balanced* in covering the major types of features in a methodology (Technical, Managerial, and Usage). The criteria's definition conforms to the Feature Analysis approach [22], in that they are of three types (based on their results): *Simple* (SM: Yes/No results), *Scale* (SC: results are discrete levels), and *Descriptive* (D: results are narrative statements).

Table 4 Criteria related to key features of web-based systems [2, 3]

Name	Type	Possible values
New or extended methodology	SM	1: New; 2: Extended (methodology)
Support for specification of technical web characteristics	SM	Yes/No
Support for architectural web design	SM	Yes/No
Support for early UI design	SM	Yes/No
Support for web-based security	SM	Yes (How?)/No
Support for rapid web development	SM	Yes (How?)/No
Support for web usability	SM	Yes (How?)/No
Addressed level of web criticality	SM	1: Low, 2: Medium, 3: High
Support for web reliability	SM	Yes (How?)/No
Support for web flexibility	SM	Yes (How?)/No
Attention to web design aspects (logic, content, navigation, UI)	SC	1: Logic; 2: Content; 3: Navigation; 4: UI; 5: Not addressed
Support for tuning the development speed based on process feedback	SC	1: No; 2: Some recommendations given; 3: Fully supported
Specification of web-related products and roles	SC	1: Only type defined; 2: Names and some recommendations given for products; 3: Fully defined

4 Results of Evaluation

The results of evaluating the targeted web development methodologies are presented in Tables 5, 6, 7 and 8, based on the type of criteria used for evaluation. If a methodology cannot be evaluated according to a certain criterion, the result has been marked with a '–'. The results clearly highlight the strengths and weaknesses of each methodology, and can be used for selecting and/or improving the methodologies.

Table 5 Results of evaluation based on general methodology evaluation criteria—Modeling language

Criterion	Targeted methodologies													
	MockupDD	RAMBUS	USABAGILE_Web	WebHelix	S-FDD	XWeb-Process	XP	UML-AW	Crystal Orange Web	S-Scrum	Scrum-CMMI	AWDWF	AWE	Scrum-MDE
Support for specific modeling language	1	1	3	1	3	1	1	3	1	1	1	1	1	3
Simplicity to learn and use	–	–	Y	–	Y	–	–	Y	–	–	–	–	–	Y
Expressiveness of modeling language	–	–	Y	–	Y	–	–	Y	–	–	–	–	–	Y
Support for complexity management	–	–	Y	–	Y	–	–	N	–	–	–	–	–	Y

Table 6 Results of evaluation based on general methodology evaluation criteria—Process

Criterion	Targeted methodologies													
	MockupDD	RAMBUS	USABAGILE_Web	WebHelix	S-FDD	XWeb-Process	XP	UML-AW	Crystal Orange Web	S-Scrum	Scrum-CMMI	AWDWF	AWE	Scrum-MDE
Coverage of generic lifecycle	D C M	D C M	D C M	D C M	D C M	D C M	D C M	D C M	-	D C M	D C M	D C M	D C M	D C M
Support for seamless transition	3	2	2	1	3	1	1	3	-	1	1	1	1	1
Support for smooth transition	3	3	3	3	3	3	3	3	-	3	3	3	3	3
Type of lifecycle	I-I.	I-I.	I-I.	I-I.	I-I.	I-I.	I-I.	I-I.	I-I.	I-I.	I-I.	I-I.	I-I.	I-I.
Attention to design activities	Y	Y	Y	Y	Y	Y	Y	Y	Y	Y	Y	Y	Y	Y
Potential of integration	1	1	1	1	1	1	1	1	2	1	1	1	1	1
Adequacy of products	3	3	3	3	3	3	3	3	-	3	3	1	3	3
Consistency of products	2	2	2	2	2	2	2	2	-	2	2	2	2	2
Support for modeling different views	S F B	S F B	S F B	S B	S F B	S	S	S F B	S	S F	-	S F B	S	S F B
Support for modeling different granularity levels	P C D	P C D O	P D	P D	S C O D F	O	O	P C O D	-	P D	P D	P D	P	P C D
Support for modeling different abstraction levels	D I	A D I	A D I	D I	A D I	D I	D I	A D I	-	A D I	A D I	A D I	I	D I
Testability of products	3	3	3	3	3	3	3	3	-	3	3	2	3	3
Tangibility of products	4	4	4	4	4	4	4	4	-	4	4	1	4	4
Traceability to requirements	Y	Y	Y	Y	Y	Y	Y	Y	-	Y	Y	Y	Y	Y
Definition of roles	1	3	1	1	3	3	3	3	3	3	3	2	3	3
Team knowledge/experience	N	Y	N	N	Y	Y	Y	Y	Y	Y	Y	Y	Y	Y
Support for team motivation	1	1	1	1	1	2	2	1	2	1	1	1	1	1
Expressiveness of process	3	2	3	3	3	3	3	3	3	3	3	3	3	2

(continued)

Table 6 (continued)

Criterion	Targeted methodologies													
	MockupDD	RAMBUS	USABAGILE_Web	WebHelix	S-FDD	XWeb-Process	XP	UML-AW	Crystal Orange Web	S-Scrum	Scrum-CMMI	AWDWF	AWE	Scrum-MDE
Completeness of definition	LAPU	LAPUR RL	LAPU	LAU	LAPU R ML	LAPURTP	LAP UR TP	LAPUR RL ML	LAUR RL	LAPU TP RRL	LAPUR RL	LAR	LAP UR TP	LAPU
Rationality and consistency	4	4	4	4	4	4	4	4	4	4	4	4	4	4
Complexity management	Y	Y	Y	Y	Y	Y	Y	Y	-	Y	Y	Y	Y	Y
Attention to detail in process	3	2	3	2	3	3	3	4	2	3	3	3	1	2
Definition of phase I/O	3	3	2	2	2	3	3	3	1	3	3	1	3	2
Availability of documentation	Y	Y	Y	Y	Y	Y	Y	Y	N	Y	Y	Y	Y	N
Tool support for process	Y	Y	N	N	Y	Y	Y	Y	-	N	N	N	N	Y
Ease of use of process	2	3	3	3	3	3	3	3	-	3	3	3	3	3
Availability of reports	Y	N	N	N	Y	Y	Y	Y	-	Y	Y	Y	Y	Y
Configurability of process	1	1	1	1	3	1	1	1	3	3	3	3	1	1
Flexibility of process	3	3	3	3	3	3	3	1	3	3	3	3	3	1
Criticality level	2	2	2	2	1	2	2	2	1	1	2	2	2	2
Platform-adaptivity of process	N	N	N	Y	N	N	N	Y	-	N	N	Y	N	N
Support for formalism	N	N	N	N	N	N	N	N	-	N	N	N	N	N
Scalability	2	2	3	3	3	2	2	2	2	3	3	2	3	2
Modularity	Y	Y	Y	Y	Y	Y	Y	Y	-	Y	Y	Y	Y	Y
Requirements elicitation	D: User Story	D: User Story	M: By Usability Experts	M: Reqs. Doc.	D: Features	D: User Story	D: User Story	M: User Reqs.	N	D: User Story	N	M: Reqs. Doc.	N	D: User Story

(continued)

Table 6 (continued)

Criterion	Targeted methodologies													
	MockupDD	RAMBUS	USABAGILE_Web	WebHelix	S-FDD	XWeb-Process	XP	UML-AW	Crystal Orange Web	S-Scrum	Scrum-CMMI	AWDWF	AWE	Scrum-MDE
Requirements specification	Mockups	Product Backlog	Prototype	Reqs. Doc.	Features	User Story	User Story	Story-board	–	Product Backlog	Product Backlog	Prototype	–	User Story
Requirements-based process	Y	Y	Y	Y	Y	Y	Y	Y	–	Y	Y	Y	Y	Y
Requirements prioritization	Y	Y	N	Y	Y	Y	Y	Y	Y	Y	Y	Y	Y	Y
Constraints/assumptions	1	1	3	1	1	2	2	1	1	2	3	3	2	1

Table 7 Results of evaluation based on agility characteristics

Criterion	Targeted methodologies													
	MockupDD	RAMBUS	USABAGILE_Web	WebHelix	S-FDD	XWebProcess	XP	UML-AW	Crystal Orange Web	S-Scrum	Scrum-CMMI	AWDWF	AWE	Scrum-MDE
Early and continuous delivery	3	3	3	3	2	3	3	2	3	3	3	3	3	3
Active user involvement	Y	Y	Y	N	Y	Y	Y	Y	Y	Y	Y	Y	Y	Y
Continuous customer feedback	Y	Y	Y	N	Y	Y	Y	Y	Y	Y	Y	Y	Y	Y
Self-organizing teams	2	1	2	1	3	3	3	3	1	3	3	1	1	3
Face-to-face conversation	Y	Y	Y	N	Y	Y	Y	Y	Y	Y	Y	Y	Y	Y
Velocity monitoring and control	Y	Y	N	N	Y	Y	Y	Y	Y	Y	Y	N	Y	Y
Team behavior/efficiency	Y	Y	Y	Y	Y	Y	Y	Y	Y	Y	Y	Y	Y	Y
Task assignment method	Team	–	–	–	Manager	Team	Team	Team	–	Team	Team	–	–	Team
Continuous integration	Y	Y	Y	Y	Y	Y	Y	Y	Y	Y	Y	Y	Y	Y
Modeling coverage	Y	Y	Y	Y	Y	Y	Y	Y	N	N	N	N	N	N
Standards	N	Y	N	N	N	Y	Y	N	N	N	Y	N	N	N
Iterative-Incremental process	Y	Y	Y	Y	Y	Y	Y	Y	Y	Y	Y	Y	Y	Y
Agile techniques	N	N	N	N	N	Y	Y	N	N	Y	Y	N	N	N
Flexibility	Y	Y	Y	Y	Y	Y	Y	Y	Y	Y	Y	Y	Y	Y
Rapid production of artifacts	2	3	2	2	3	3	3	2	3	3	3	3	3	3
Leanness	Y	Y	Y	Y	Y	Y	Y	Y	Y	Y	Y	Y	Y	Y
Learning	2	3	3	2	3	3	3	3	3	3	3	3	3	3
Feedback by process	Y	Y	Y	Y	Y	Y	Y	Y	Y	Y	Y	Y	Y	Y

Table 8 Results of evaluation based on key features of web-based systems

Criterion	Targeted methodologies													
	Mockup-DD	RAMBUS	USABAGILE_Web	WebHelix	S-FDD	XWeb-Process	XP	UML-AW	Crystal Orange Web	S-Scrum	Scrum-CMMI	AWDWF	AWE	Scrum-MDE
New or extended methodology	1	1	2	1	2	2	1	2	1	2	2	2	1	2
Specification of technical web characteristics	N	N	N	N	N	N	N	N	N	N	N	N	Y	N
Architectural web design	Y	Y	N	Y	Y	Y	Y	Y	Y	Y	Y	Y	Y	Y
Early UI design	Y	Y	Y	Y	Y	Y	Y	Y	Y	Y	Y	Y	Y	Y
Web-based security	Y	Y	Y	Y	Y	Y	Y	Y	Y	Y	Y	Y	Y	Y
Rapid web development	Y	Y	Y	Y	Y	Y	Y	Y	Y	Y	Y	Y	Y	Y
Web usability	Y	Y	Y	Y	Y	Y	Y	Y	Y	Y	Y	Y	Y	Y
Addressed level of web criticality	Y	Y	Y	Y	Y	Y	Y	Y	Y	Y	Y	Y	Y	Y
Web reliability	Y	Y	Y	Y	Y	Y	Y	Y	Y	Y	Y	Y	Y	Y
Web flexibility	Y	Y	Y	Y	Y	Y	Y	Y	Y	Y	Y	Y	Y	Y
Attention to web design aspects	1 2 3 4	1 2 3 4	1 2 3 4	1 2 3 4	1 2 3 4	1 2 3 4	1 4	1 2 3 4	1 4	1 2 3 4	1 2 3 4	1 2 3 4	1 2 3 4	1 2 3 4
Tunability of development speed	2	3	2	2	2	3	3	2	3	2	2	2	2	2
Specification of web-related products and roles	2	3	1	2	3	3	3	3	3	3	3	1	3	3

5 Conclusions and Future Work

By evaluating the targeted methodologies, their individual strengths and weaknesses are highlighted. However, apart from these evaluations, some general observations can also be made: it can be observed that some of the targeted methodologies pay special attention to web security issues, a feature that is increasingly considered as essential in modern web systems; it can also be observed that Scrum variants seem to fully cover the different web development contexts that are commonly encountered.

As future work, we intend to propose a comprehensive agile web development methodology that addresses the weaknesses of existing methodologies while making use of their strengths. Another strand of research can focus on using the evaluation results for extending existing methodologies so that their shortcomings are properly addressed.

References

1. Farahani, F.F., Ramsin, R.: Methodologies for agile product line engineering: a survey and evaluation. In: Proceedings of the International Conference on Intelligent Software Methodologies, Tools and Techniques (SOMET'14), pp. 545–564 (2014)
2. Babanezhad, R., Bibalan, Y.M., Ramsin, R.: Process patterns for web engineering. In: Proceedings of the Computer Software and Applications Conference (COMPSAC'10), pp. 477–486 (2010)
3. Kaur, S., Singh, H.: Quality metrics for agile web engineering based on GQM approach. VSRD-IJCSIT 2(6), 454–461 (2012)
4. Rivero, J.M., et al.: Mockup-driven development: providing agile support for model-driven web engineering. Inf. Softw. Technol. 56(6), 670–687 (2014)
5. Pereira, V., Francisco, A.: Introducing a new agile development for web applications using a groupware as example. Commun. Comput. Inf. Sci. 165, 144–160 (2011)
6. Benigni, G., Gervasi, O., Passeri, F.L., Kim, T.: USABAGILE_Web: a web agile usability approach for web site design. Lect. Notes Comput. Sci. (LNCS) 6017, 422–431 (2010)
7. Whitson, G.: WebHelix: another web engineering process. J. Comput. Sci. Coll. 21(5), 21–27 (2006)
8. Subramanian, N., Whitson, G.: Augmented WebHelix: a practical process for web engineering. In: Software Engineering for Modern Web Applications: Methodologies and Technologies, pp. 25–27. IGI Global (2008)
9. Ge, X., et al.: Agile development of secure web applications. In: Proceedings of the International Conference on Web Engineering (ICWE'06), pp. 305–312 (2006)
10. Sampaio, A., Vasconcelos, A., Sampaio, P.R.F.: Design and empirical evaluation of an agile web engineering process. In: Proceedings of the Brazilian Symposium on Software Engineering (SBES'04), pp. 194–209 (2004)
11. Maurer, F., Martel, S.: Extreme programming: rapid development for web-based applications. Internet Comput. 6(1), 86–91 (2002)
12. Ambler, S.W.: AM Throughout the XP Lifecycle. http://www.agilemodeling.com/essays/agileModelingXPLifecycle.htm (2002)

13. Lee, W., et al.: Agile development of web application by supporting process execution and extended UML model. In: Proceedings of the Asia-Pacific Software Engineering Conference (APSEC'05), pp. 93–200 (2005)
14. Cockburn, A.: Agile Software Development: The Cooperative Game. Addison-Wesley (2002)
15. Mougouei, D., Fazlida, N., Sani, M., Almasi, M.M.: S-Scrum: a secure methodology for agile development of web services. WCSIT J. 3(1), 15–19 (2013)
16. Salinas, C.J.T., Escalona, M.J., Mejías, M.: A scrum-based approach to CMMI maturity level 2 in web development environments. In: Proceedings of the International Conference on Information Integration and Web-based Applications and Services (IIWAS'12), pp. 282–285 (2012)
17. Hu, R., Wang, Z., Hu, J., Xu, J., Xie, J.: Agile web development with web framework. In: Proceedings of the International Conference on Wireless Communications, Networking and Mobile Computing (WiCOM'08), pp. 1–4 (2008)
18. McDonald, A., Welland, R.: Agile Web Engineering (AWE) process: multidisciplinary stakeholders and team communication. In: Proceedings of the International Conference on Web Engineering (ICWE'03), pp. 515–518 (2003)
19. Basso, F.P., Pillat, R.M., Roos-Frantz, F., Frantz, R.Z.: Study on combining model-driven engineering and Scrum to produce web information systems. In: Proceedings of the International Conference on Enterprise Information Systems (ICEIS'14), pp. 137–144 (2014)
20. Hesari, S., Mashayekhi, H., Ramsin, R.: Towards a general framework for evaluating software development methodologies. In: Proceedings of the Computer Software and Applications Conference (COMPSAC'10), pp. 208–217 (2010)
21. Karam, G.M., Casselman, R.S.: A cataloging framework for software development methods. Computer 26(2), 34–44 (1993)
22. Kitchenham, B., Linkman, S., Law, D.: DESMET: a methodology for evaluating software engineering methods and tools. Comput. Control Eng. J. 8(3), 120–126 (1997)

Load Experiment of the vDACS Scheme in Case of Increasing the Simultaneous Connection for the DACS SV

Kazuya Odagiri, Shogo Shimizu and Naohiro Ishii

Abstract In the current Internet system, there are many problems using anonymity of the network communication such as personal information leaks and crimes using the Internet system. This is why TCP/IP protocol used in Internet system does not have the user identification information on the communication data, and it is difficult to supervise the user performing the above acts immediately. As a study for solving the above problem, there is the study of Policy Based Network Management (PBNM). This is the scheme for managing a whole Local Area Network (LAN) through communication control for every user. In this PBNM, two types of schemes exist. The first is the scheme for managing the whole LAN by locating the communication control mechanisms on the path between network servers and clients. The second is the scheme of managing the whole LAN by locating the communication control mechanisms on clients. As the second scheme, we have studied theoretically about the Destination Addressing Control System (DACS) Scheme. By applying this DACS Scheme to Internet system management, we will realize the policy-based Internet system management. In this paper, as the progression phase for the last goal, we perform the load experiment of the cloud type virtual PBNM named the vDACS Scheme, which can be used by plural organizations, for applications to the small and medium size scale organization. The number of clients used in an experiment is 200.

Keywords Policy-based network management · DACS scheme · NAPT

K. Odagiri (✉)
Sugiyama Jogakuen University, 17-3 Hosigaokamotomachi Chiksa-ku,
Nagoya, Aichi 464-8662, Japan
e-mail: kodagiri@sugiyama-u.ac.jp; kazuodagiri@yahoo.co.jp

S. Shimizu
Gakushuin Women's College, Tokyo, Japan
e-mail: shogo.shimizu@gakushuin.ac.jp

N. Ishii
Aichi Institute of Technology, Toyota, Aichi, Japan
e-mail: ishii@aitech.ac.jp

© Springer International Publishing AG 2018
R. Lee (ed.), *Software Engineering Research, Management and Applications*,
Studies in Computational Intelligence 722, DOI 10.1007/978-3-319-61388-8_2

1   Introduction

In the current Internet system, there are many problems using anonymity of the network communication such as personal information leaks and crimes using the Internet system. The news of the information leak in the big company is sometimes reported through the mass media. Because TCP/IP protocol used in Internet system does not have the user identification information on the communication data, it is difficult to supervise the user's acts above immediately. As studies and technologies for managing Internet system realized on TCP/IP protocol, those such as Domain Name System (DNS), Routing protocol, Fire Wall (F/W) and Network address port translation (NAPT)/network address translation (NAT) are listed. Except these studies, various studies are performed elsewhere. However, they are the studies for managing the specific part of the Internet system, and have no purpose of solving the above problems.

As a study for solving the problems, Policy Based Network Management (PBNM) [1] exists. The PBNM is a scheme for managing a whole Local Area Network (LAN) through communication control every user, and cannot be applied to the Internet system. This PBNM is often used in a scene of campus network management. In a campus network, network management is quite complicated. Because a computer management section manages only a small portion of the wide needs of the campus network, there are some user support problems. For example, when mail boxes on one server are divided and relocated to some different server machines, it is necessary for some users to update a client machine's setups. Most of computer network users in a campus are students. Because students do not check frequently their e-mail, it is hard work to make them aware of the settings update. This administrative operation is executed by means of web pages and/or posters. For the system administrator, individual technical support is a stiff part of the network management. Because the PBNM manages a whole LAN, it is easy to solve this kind of problem. In addition, for the problem such as personal information leak, the PBNM can manage a whole LAN by making anonymous communication non-anonymous. As the result, it becomes possible to identify the user who steals personal information and commits a crime swiftly and easily. Therefore, by applying the PBNM, we will study about the policy-based Internet system management.

In the existing PBNM, there are two types of schemes. The first is the scheme of managing the whole LAN by locating the communication control mechanisms on the path between network servers and clients. The second is the scheme of managing the whole LAN by locating the communication control mechanisms on clients. It is difficult to apply the first scheme to Internet system management practically, because the communication control mechanism needs to be located on the path between network servers and clients without exception. Because the second scheme locates the communication control mechanisms as the software on each client, it becomes possible to apply the second scheme to Internet system management by devising the installing mechanism so that users can install the software to the client easily.

As to the second scheme, we have studied theoretically about the Destination Addressing Control System (DACS) Scheme. As the works on the DACS Scheme, we showed the basic principle of the DACS Scheme, and security function [2]. After that, we implemented a DACS System to realize a concept of the DACS Scheme. By applying this DACS Scheme to Internet system, we will realize the policy-based Internet system management. Then, the Wide Area DACS system (wDACS system) [3] to use it in one organization was showed as the second phase for the last goal. As the first step of the second phase, we showed the concept of the cloud type virtual PBNM, which could be used by plural organizations [4]. In this paper, as the progression phase of the third phase for the last goal, we perform the load experiment to confirm the possibility of the cloud type virtual PBNM for the use in plural organizations. In Sect. 2, motivation and related research for this study are described. In Sect. 3, the existing DACS Scheme and wDACS Scheme is described. In Sect. 4 the proposed scheme and load experiment results are described.

2 Motivation and Related Reserach

In the current Internet system, problems using anonymity of the network communication such as personal information leak and crimes using the Internet system occur. Because TCP/IP protocol used in Internet system does not have the user identification information on the communication data, it is difficult to supervise the user performing the above acts immediately.

As studies and technologies for Internet system management to be comprises of TCP/IP [5], many technologies are studied. For examples, Domain name system (DNS), Routing protocol such as Interior gateway protocol (IGP) such as Routing information protocol (RIP) and Open shortest path first (OSPF), Fire Wall (F/W), Network address translation (NAT)/Network address port translation (NAPT), Load balancing, Virtual private network (VPN), Public key infrastructure (PKI), Server virtualization. Except these studies, various studies are performed elsewhere. However, they are for managing the specific part of the Internet system, and have no purpose of solving the above problems.

As a study for solving the above problem, the study area about PBNM exists. This is a scheme of managing a whole LAN through communication control every user. Because this PBNM manages a whole LAN by making anonymous communication non-anonymous, it becomes possible to identify the user who steals personal information and commits a crime swiftly and easily. Therefore, by applying this policy-based thinking, we study about the policy-based Internet system management.

In policy-based network management, there are two types of schemes. The first scheme is the scheme described in Fig. 1. The standardization of this scheme is

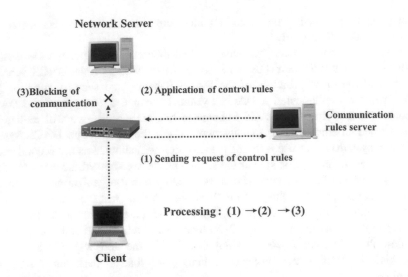

Network Server

(3)Blocking of
communication ✕

(2) Application of control rules

Communication
rules server

(1) Sending request of control rules

Processing : (1) →(2) →(3)

Client

Fig. 1 Principle in first scheme

performed in various organizations. In IETF, a framework of PBNM [1] was established. Standards about each element constituting this framework are as follows. As a model of control information stored in the server called Policy Repository, Policy Core Information model (PCIM) [6] was established. After it, PCMIe [7] was established by extending the PCIM. To describe them in the form of Lightweight Directory Access Protocol (LDAP), Policy Core LDAP Schema (PCLS) [8] was established. As a protocol to distribute the control information stored in Policy Repository or decision result from the PDP to the PEP, Common Open Policy Service (COPS) [9] was established. Based on the difference in distribution method, COPS usage for RSVP (COPS-RSVP) [10] and COPS usage for Provisioning (COPS-PR) [11] were established. RSVP is an abbreviation for Resource Reservation Protocol. The COPS-RSVP is the method as follows. After the PEP having detected the communication from a user or a client application, the PDP makes a judgmental decision for it. The decision is sent and applied to the PEP, and the PEP adds the control to it. The COPS-PR is the method of distributing the control information or decision result to the PEP before accepting the communication.

Next, in DMTF, a framework of PBNM called Directory-enabled Network (DEN) was established. Like the IETF framework, control information is stored in the server storing control information called Policy Server, which is built by using the directory service such as LDAP [12], and is distributed to network servers and networking equipment such as switch and router. As the result, the whole LAN is managed. The model of control information used in DEN is called Common Information Model (CIM), the schema of the CIM (CIM Schema Version 2.30.0) [13] was opened. The CIM was extended to support the DEN [14], and was incorporated in the framework of DEN.

Fig. 2 Essential principle

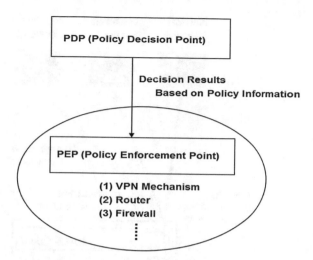

In addition, Resource and Admission Control Subsystem (RACS) [15] was established in Telecoms and Internet converged Services and protocols for Advanced Network (TISPAN) of European Telecommunications Standards Institute (ETSI), and Resource and Admission Control Functions (RACF) was established in International Telecommunication Union Telecommunication Standardization Sector (ITU-T) [16].

However, all the frameworks explained above are based on the principle shown in Fig. 1. As problems of these frameworks, two points are presented as follows. Essential principle is described in Fig. 2. To be concrete, in the point called PDP (Policy Decision Point), judgment such as permission or non-permission for communication pass is performed based on policy information. The judgment is notified and transmitted to the point called the PEP, which is the mechanism such as VPN mechanism, router and Fire Wall located on the network path among hosts such as servers and clients. Based on that judgment, the control is added for the communication that is going to pass by.

The principle of the second scheme is described in Fig. 3. By locating the communication control mechanisms on the clients, the whole LAN is managed. Because this scheme controls the network communications on each client, the processing load is low. However, because the communication control mechanisms needs to be located on each client, the work load becomes heavy.

When it is thought that Internet system is managed by using these two schemes, it is difficult to apply the first scheme to Internet system management practically. This is why the communication control mechanism needs to be located on the path between network servers and clients without exception. On the other hand, the second scheme locates the communication controls mechanisms on each client. That is, the software for communication control is installed on each client. So, by devising the installing mechanism and letting users install software to the client easily, it becomes possible to apply the second scheme to Internet system

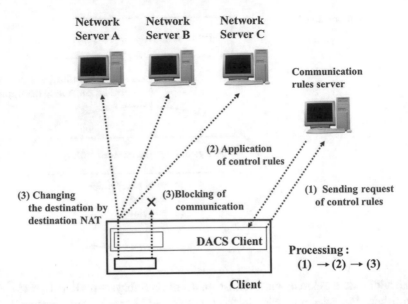

Fig. 3 Principle in second scheme

management. As a first step for the last goal, we showed the Wide Area DACS system (wDACS) system [3]. This system manages a wide area network, which one organization manages. Therefore, it is impossible for plural organizations to use this system. Then, as the first step of the second phase, we showed the concept of the cloud type virtual PBNM, which could be used by plural organizations in this paper.

3   Existing DACS Scheme and wDACS System

In this section, the content of the DACS Scheme which is the study of the phase 1 is described.

3.1   Basic Principle of the DACS Scheme

Figure 4 shows the basic principle of the network services by the DACS Scheme. At the timing of the (a) or (b) as shown in the following, the DACS rules (rules defined by the user unit) are distributed from the DACS Server to the DACS Client.

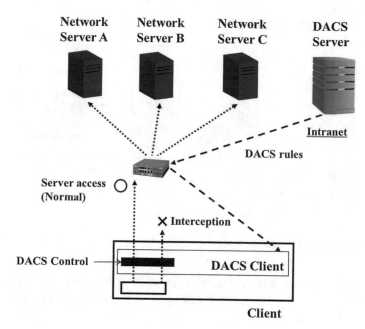

Fig. 4 Basic principle of the DACS scheme

(a) At the time of a user logging in the client.
(b) At the time of a delivery indication from the system administrator.

According to the distributed DACS rules, the DACS Client performs (1) or (2) operation are shown in the following. Then, communication control of the client is performed for every login user.

(1) Destination information on IP Packet, which is sent from application program, is changed.
(2) IP Packet from the client, which is sent from the application program to the outside of the client, is blocked.

An example of the case (1) is shown in Fig. 4. In Fig. 4, the system administrator can distribute a communication of the login user to the specified server among servers A, B or C. Moreover, the case (2) is described. For example, when the system administrator wants to forbid a user to use MUA (Mail User Agent), it will be performed by blocking IP Packet with the specific destination information.

In order to realize the DACS Scheme, the operation is done by a DACS Protocol as shown in Fig. 5. As shown by (1) in Fig. 5, the distribution of the DACS rules is performed on communication between the DACS Server and the DACS Client, which is arranged at the application layer. The application of the DACS rules to the DACS Control is shown by (2) in Fig. 5.

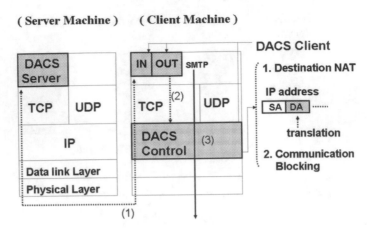

Fig. 5 Layer setting of the DACS scheme

The steady communication control, such as a modification of the destination information or the communication blocking is performed at the network layer as shown by (3) in Fig. 5.

3.2 Communication Control on Client

The communication control on every user was given. However, it may be better to perform communication control on every client instead of every user. For example, it is the case where many unspecified users use a computer room, which is controlled. In this section, the method of communication control on every client is described, and the coexistence method with the communication control on every user is considered.

When a user logs into a client, the IP address of the client is transmitted to the DACS Server from the DACS Client. Then, if the DACS rules corresponding to IP address, is registered into the DACS Server side, it is transmitted to the DACS Client. Then, communication control for every client can be realized by applying the DACS Control. In this case, it is a premise that a client uses a fixed IP address. However, when using DHCP service, it is possible to carry out the same control to all the clients linked to the whole network or its subnetwork for example.

When using communication control on every user and every client, communication control may conflict. In that case, a priority needs to be given. The judgment is performed in the DACS Server side as shown in Fig. 6. Although not necessarily stipulated, the network policy or security policy exists in the organization such as a university (1). The priority is decided according to the policy (2). In (a), priority is

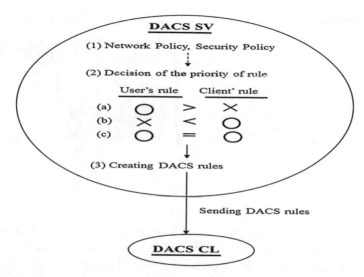

Fig. 6 Creating the DACS rules on the DACS server

given for the user's rule to control communication by the user unit. In (b), priority is given for the client's rule to control communication by the client unit. In (c), the user's rule is the same as the client's rule. As the result of comparing the conflict rules, one rule is determined respectively. Those rules and other rules not over-lapping are gathered, and the DACS rules are created (3). The DACS rules are transmitted to the DACS Client. In the DACS Client side, the DACS rules are applied to the DACS Control. The difference between the user's rule and the client's rule is not distinguished.

3.3 Security Mechanism of the DACS Scheme

In this section, the security function of the DACS Scheme is described. The communication is tunneled and encrypted by use of SSH. By using the function of port forwarding of SSH, it is realized to tunnel and encrypt the communication between the network server and the, which DACS Client is installed in. Normally, to communicate from a client application to a network server by using the function of port forwarding of SSH, local host (127.0.0.1) needs to be indicated on that client application as a communicating server. The transparent use of a client, which is a characteristic of the DACS Scheme, is failed. The transparent use of a client means

Fig. 7 Extend security
function

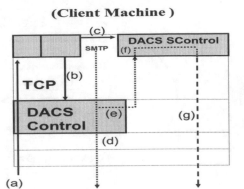

that a client can be used continuously without changing setups when the network system is updated. The function that doesn't fail the transparent use of a client is needed. The mechanism of that function is shown in Fig. 7.

3.4 Application to Cloud Environment

In this section, the contents of wDACS system are explained in Fig. 8. First, as preconditions, because private IP addresses are assigned to all servers and clients existing in from LAN1 to LAN n, mechanisms of NAT/NAPT are necessary for the communication from each LAN to the outside. In this case, NAT/NAPT is located on the entrance of the LAN such as (1), and the private IP address is converted to the global IP address towards the direction of the arrow. Next, because the private IP addresses are set on the servers and clients in the LAN, other communications except those converted by Destination NAT cannot enter into the LAN. But, responses for the communications sent from the inside of the LAN can enter into the inside of the LAN because of the reverse conversion process by the NAT/NAPT. In addition, communications from the outside of the LAN1 to the inside are performed through the conversion of the destination IP address by Destination NAT. To be concrete, the global IP address at the same of the outside interface of the router is changed to the private IP address of each server. From here, system configuration of each LAN is described. First, the DACS Server and the authentication server are located on the DMZ on the LAN1 such as (4). On the entrance of the LAN1, NAT/NAPT and destination NAT exists such as (1) and (2). Because only the DACS Server and network servers are set as the target destination, the authentication server cannot be accessed from the outside of the LAN1. In the LANs form LAN 2 to LAN n, clients managed by the wDACS system exist, and NAT/NAPT is located on the entrance of each LAN such as (1). Then, F/W such as (3) or (5) exists behind or with NAT/NAPT in all LANs.

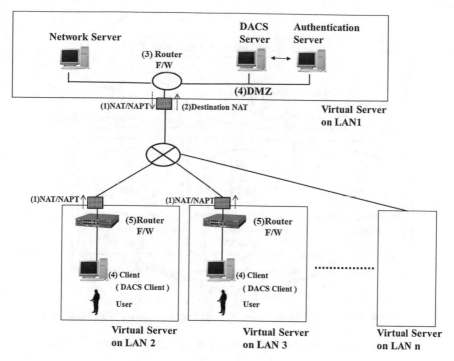

Fig. 8 Basic system configuration of wDACS system

4 Cloud Type Virtual PBNM for the Common Use Between Plural Organizations

In this section, after the concept and implementation of the proposed scheme were described, functional evaluation results are described.

4.1 Concept of the Cloud Type Virtual PBNM for the Common Use Between Plural Organizations

In Fig. 9 which is described in [4], the proposed concept is shown. Because the existing wDACS Scheme realized the PBNM control with the software called the DACS Server and the DACS client, other mechanism was not needed. By this point, application to the cloud environment was easy.

The proposed scheme in this paper realizes the common usage by plural organizations by adding the following elements to realize the common usage by plural organizations: user identification of the plural organizations, management of the policy information of the plural organizations, application of the PKI for code

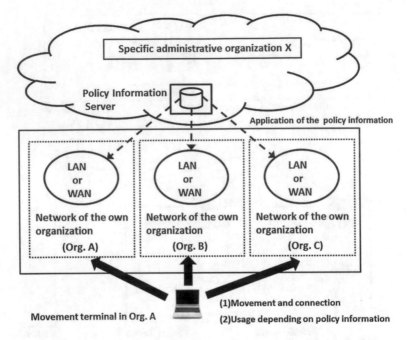

Fig. 9 Basic system configuration of wDACS system concept of the proposed scheme

communication in the Internet, Redundant configuration of the DACS Server (policy information server), load balancing configuration of the DACS Server, installation function of DACS Client by way of the Internet.

4.2 Implementation of the Basic Function in the Cloud Type Virtual PBNM for the Common Usage Between Plural Organizations

In the past study [2], the DACS Client was operated on the windows operation system (Windows OS). It was because there were many cases that the Windows OS was used as the client. However, the Linux operating system (Linux OS) had enough functions to be used as the client recently, too. In addition, it was thought that the case used in the clients in the future came out recently. Therefore, to prove the possibility of the DACS Scheme on the Linux OS, the basic function of the DACS Client was implemented in this study. The basic functions of the DACS Server and DACS Client were implemented by JAVA language. From here, it is described about the order of the process in the DACS Client and DACS Server as follows.

(Processes in the DACS Client)
(p1) The information acquisition from Cent OS
(p2) Transmission from the DACS Client to the DACS
(p3) The information transmission from the DACS Client to
(p4) The reception of the DACS rules from the DACS Server
(p5) Application of the DACS rules of the DACS Control

(Processes in the DACS Server)
(p1) The information reception from the DACS Client
(p2) Connection to the database
(p3) Inquiry of the Database
(p4) Transmission of the DACS rules to the DACS Client.

4.3   Results of the Functional Evaluation

In this section, the results of the functional evaluation for the implementation system are described in Fig. 10.

In Fig. 11, the setting situation of the DACS rules is described. This DACS rules is the rule to change a Web server for the access. The delivery of the DACS rules is between the DACS SV and the DACS CL encrypted by using SSL.

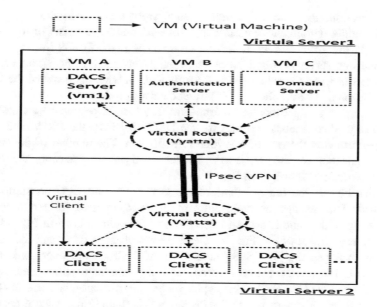

Fig. 10 Prototype system

```
<?XML version="1.0" encoding="uft8"?>

<direct>

  <rule priority="0" table="nat" ipv="ipv4" chain="PREROUTING_direct">
-d 192.168.1.10:80 -j DNAT --to 192.168.1.12:80</rule>

</direct>
```

Fig. 11 Setting situation of the DACS rules on the DACS CL

By this DACS rules, the next operation was realized. When the user accessed the Web Server with the IP address of 192.168.1.10, the Web Server with the IP address of 192.168.1.12 was accessed actually. As for this communication result, the communication log on each Web server was confirmed by viewing.

5 Load Experiment Results

5.1 Load Experiment Results to Confirm the Function of the Software for Realization of the Cloud Type Virtual PBNM for the Common Use Between Plural Organizations

In this section, the load experiment results are described. In the Fig. 12, the experimental environment is described. This environment consists of four virtual servers. In the virtual server 1, servers group such as the DACS Server and user authentication server is stored. In other virtual severs such as the virtual server 2, virtual server 3 and virtual server 4, the virtual client which is installed the DACS Client is stored. The number of the virtual clients is 100.

By using this experimental environment, the load experiment was executed. Specifically, simultaneous accesses for the DACS SV from the 100 virtual clients were performed at the rate of one time form 15 min. The number of the simultaneous connection for the DACS SV was set to 10 on this occasion. The experimental results are described in Fig. 13.

In this Figure, the practice time of the DACL CL and CPU consumption is described. The average of the results of the measurement for ten times was 263.2 MHz. This value is around three times of the value shown in Fig. 14.

In the experiment of the Fig. 14, the Windows client is used, and the communications between the DACS SV and the DACS CL is not encrypted. In this experiment, the Linux client is used, and the communications between the DACS SV and the DACS CL is encrypted by SSL. Particularly, because an element of the overhead processes of the SSL is large, it is thought that such a result was derived.

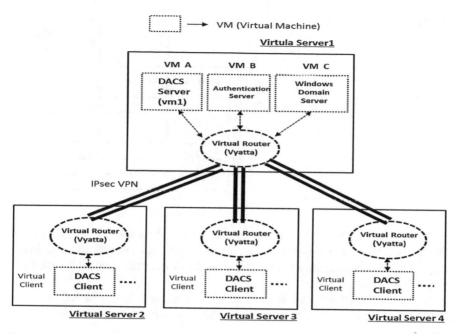

Fig. 12 Experimental environment

	Practice time	CPU Consumption
1	2:02	268
2	2:18	268
3	2:33	276
4	2:48	251
5	3:03	261
6	3:18	252
7	3:33	248
8	3:48	265
9	4:03	265
10	4:18	278

Fig. 13 Experimental results (1)

5.2 Load Experiment Results for Applications to the Small and Medium Size Scale Organization

In this section, the load experiment results are described. The experimental environment is described. The experimental environment is as previous experiment environment. The simultaneous accesses for the DACS SV from the 200 virtual clients were performed at the rate of one time form 15 min.

Fig. 14 Experimental results
(2)

	CPU consumption
1	59
2	58
3	51
4	58
5	59
6	59
7	53
8	51
9	53
10	58

In the first experiment, the number of the simultaneous connection for the DACS SV was set to 10 on this occasion. The experimental results are described in Fig. 15. In this Figure, the average of the results of the measurement for ten times was 477.3 MHz.

In the second experiment, the number of the simultaneous connection for the DACS SV was set to 20 on this occasion. The experimental results are described in Fig. 16.

In this Figure, the average of the results of the measurement for ten times was 538.8 MHz. Then, in the Fig. 17, the load experiment results are described in the case that the number of the simultaneous connection for the DACS SV was set to 30 on. As the result, the average of the CPU consumption was 540.1 MHz.

As the results of these experiments, the CPU consumption becomes approximately constant in the case of the number of the simultaneous connection from 20 to 30. The value of around 540 MHz is the CPU load when the 200 clients are connected simultaneously.

	Practice Time	CPU Consumption
1	16:45	490
2	17:00	491
3	17:15	474
4	17:30	487
5	17:45	471
6	18:00	478
7	18:15	466
8	18:30	473
9	18:45	474
10	19:00	469

Fig. 15 Experimental results (3)

	Practice Time	CPU Consumption
1	20:00	567
2	20:15	560
3	20:30	538
4	20:45	537
5	21:00	530
6	21:15	525
7	21:30	538
8	21:45	530
9	22:00	525
10	22:15	538

Fig. 16 Experimental results (4)

	Practice Time	CPU Consumption
1	23:15	564
2	23:30	567
3	23:45	536
4	0:00	546
5	0:15	520
6	0:30	543
7	0:45	531
8	1:00	515
9	1:15	543
10	1:30	536

Fig. 17 Experimental results (5)

6 Conclusion

In this paper, we performed the load experiment of the cloud type virtual PBNM, which can be used by plural organizations. In this experiment, the 200 virtual clients with Linux OS are used, and the communications between the DACS SV and the DACS CL are encrypted. The number of the simultaneous connection for the DACS SV was set to 20 on this occasion. As the result, the average of CPU consumption was 538.8 MHz. When the number of the simultaneous connection for the DACS SV was set to 30 on this occasion, the average of CPU consumption was 540.1 MHz. These two values are two times as large as the value in case of the 10 simultaneous connections.

As to the future work, we are going to perform more load experiments in the form of increasing the number of the virtual client and the number of the simultaneous connection for the DACS SV.

Acknowledgements This work was supported by the research grant of KDDI Foundation. We express our gratitude.

References

1. Yavatkar, R., Pendarakis, D., Guerin, R.: A Framework for Policy-Based Admission Control. IETF RFC 2753 (2000)
2. Odagiri, K., Yaegashi, R., Tadauchi, M., Ishii, N.: Secure DACS scheme. J. Netw. Comput. Appl., Elsevier **31**(4), 851–861 (2008)
3. Odagiri, K., Shimizu, S., Takizawa, M., Ishii, N.: Theoretical suggestion of policy-based wide area network management system (wDACS system part-I). Int. J. Netw. Distrib. Comput. (IJNDC), **1**(4), 260–269 (2013)
4. Odagiri, K., Shimizu, S., Ishii, N., Takizawa, M.: Suggestion of the cloud type virtual policy based network management scheme for the common use between plural organizations. In: Proceedings of International Conference on Network-Based Information Systems (NBiS-2015), September, pp. 180–186 (2015)
5. Cerf, V., Kahn, E.: A protocol for packet network interconnection. IEEE Trans. on Commun. **COM-22**, 637–648 (1974)
6. Moore, B., et al.: Policy Core Information Model—Version 1 Specification. IETF RFC 3060 (2001)
7. Moore, B.: Policy Core Information Model (PCIM) Extensions. IETF 3460 (2003)
8. Strassner, J., Moore, B., Moats, R., Ellesson, E.: Policy Core Lightweight Directory Access Protocol (LDAP) Schema. IETF RFC 3703 (2004)
9. Durham, D., et al.: The COPS (Common Open Policy Service) Protocol. IETF RFC 2748 (2000)
10. Herzog, S., et al.: COPS usage for RSVP. IETF RFC 2749 (2000)
11. Chan, K., et al.: COPS Usage for Policy Provisioning (COPS-PR). IETF RFC 3084 (2001)
12. CIM Core Model V2.5 LDAP Mapping Specification (2002)
13. CIM Schema: Version 2.30.0 (2011)
14. Wahl, M., Howes, T., Kille, S.: Lightweight Directory Access Protocol (v3). IETF RFC 2251 (1997)
15. ETSI ES 282 003: Telecoms and Internet converged Services and protocols for Advanced Network (TISPAN). Resource and Admission Control Subsystem (RACS). Functional Architecture, June 2006
16. ETSI ETSI ES 283 026: Telecommunications and Internet Converged Services and Protocols for Advanced Networking (TISPAN). Resource and Admission Control. Protocol for QoS reservation information exchange between the Service Policy Decision Function (SPDF) and the Access-Resource and Admission Control Function (A-RACF) in the Resource and Protocol specification, April 2006

Blind Channel Estimation Using Novel Independent Component Analysis with Pulse Shaping for Interference Cancellation

Renuka Bhandari and Sangeeta Jadhav

Abstract Now days with the growing exposure of wireless communications, there is more focus on achieving the spectral efficiency and low bit rate errors (BER). This can be basically achieved by Space Time Frequency based Multiple Input Multiple Output (MIMO)-OFDM wireless systems. The efficient channel estimation method plays important role in optimizing the performance of spectral efficiency and BER. There are different types of MIMO-OFDM channel estimation methods. In this paper, we focused on designing efficient blind channel estimation method for MIMO-OFDM. Recently there has been increasing research interest in designing the blind channel based estimation methods. There are number of blind channel estimation methods introduced so far, however none of them effectively addressed the problem of Inter Symbol Interference (ISI). ISI may have worst impact on performance of channel estimation methods if there are not addressed by channel estimation techniques. In this paper we are designing the novel blind channel estimation approach using Independent Component Analysis (ICA) with both ISI cancellation and blind interference cancellation. This method is named as Hybrid ICA (HICA). HICA algorithm use the HOS (higher order statistical) approach and pulse shaping in order to minimize the blind interference and ISI effects. Simulation results shows that HICA is outperforming the existing channel estimation methods in terms of BER and MSE.

Keywords MIMO-OFDM · Channel estimation · Spectral efficiency · Error rates · ICA · Interference

R. Bhandari (✉)
Department of E&TC, Dr. D.Y. Patil Institute of Engineering & Technology,
Pune 411018, India
e-mail: renukabhandari6@gmail.com

R. Bhandari · S. Jadhav
Army Institute of Technology Pune, Pune 411015, India

© Springer International Publishing AG 2018
R. Lee (ed.), *Software Engineering Research, Management and Applications*,
Studies in Computational Intelligence 722, DOI 10.1007/978-3-319-61388-8_3

1 Introduction

In wireless communications, the approach multiple input multiple output (MIMO) is widely used in which there are multiple antennas at sender as well as receiver side. MIMO is nothing but the vital approach designed for current generation of digital/wireless communication standards. The MIMO was integrated with communication or transmission systems such as OFDM (orthogonal frequency division multiplexing) as well as CDMA (code division multiple access) [1]. MIMO-OFDM is our main focus in this paper. MIMO-OFDM transmission methods are widely studied since from last one decade. In MIMO, basically the transmitter antennas are employed in order to gain the higher data rates using spatial multiplexing and optimize the link reliability using either of three coding standards such as (1) space-time, (2) space-frequency and (3) space-time-frequency. The basic characteristic of all three coding standards is assumption of accurate channel information at the side of receiver. In case of practice, when the channel information is not available, design of receiver is basically depends on the suboptimum equalization differentiation solutions in order to track and acquire the data at receiver using training sequence. But the training sequence is leads to be overhead limitation which may be prohibitive [2].

Now days, the mobile wireless systems are demanded for more data rate for different multimedia services. Therefore MIMO-OFDM transmission systems are now considered as strong methodology for designing the wireless communication systems. This is because of distinct benefits of both OFDM and MIMO. The channel estimation methods of MIMO-OFDM are categorized in three main categories such as training based, semi blind and blind channel estimation methods. In first category, prepare known training samples in order to perform the proper channel estimation. The LS (least square) and MMSE (minimum mean square error) are the well known examples of training based channel estimation methods. The second category is based on combined properties of training based and blind based channel estimation methods and used with MIMO-OFDM communication systems. Third channel estimation approach is called as blind method in which second order stationary statistics (SOS) or higher order statistical (HOS) are used for delivering higher spectral efficiency. In wireless communication systems, the wireless channel frequently designed as sparse channel with the higher delay spread [3], however number of significant non zero paths basically very small. Depending on assumption of sparsity of equivalent discrete-time channel in which only some taps in line of longer tapped delay are significantly considered. In CDMA and OFDM systems, sparse structure of wireless channel is widely used in order to optimize the channel estimation performance [4]. There are different sparse channel based estimation techniques which are utilizing the training sequence and command with two main steps such as (1) position detection of MSTs (most significant taps), also called as non-zero taps, (2) effective channels estimation fetching by using the MSTs position.

The scope and goal of this paper is to present the novel approach of blind channel estimation method for MIMO-OFDM. There is increasing research interest in designing blind channel estimation techniques. There are number of recent blind channel estimation techniques claims that increasing researchers interest [1–3]. There are number reasons due which there is increasing researches on blind channel estimation approaches. In OFDM systems, basically symbols are transmitted in the form of blocks, therefore approach of iterative channel estimation and block based is enabled in 4th generation wireless systems. Therefore, first obtain the initial symbol estimates by using the blind channel estimation method and then employ the preliminary symbol estimations to gain the higher fidelity channel estimation. This process iteratively repeated with the soft information exchange in order improves the both data symbol estimations as well as channel estimates [5]. Another motivation of increasing research studies on blind channel estimations is the present architecture of heterogeneous wireless systems with small cells like femto cells those are having low mobility based users. This leads in rapid increase in low mobility based applications and hence motivating to design blind channel estimation which basically needs the large number of samples for good performance based on quasi static channel conditions.

Therefore numbers of blind channel estimation based methods are designed in previous studies. In [4], blind channel estimation based on CP (cyclic prefix) based redundancy. This approach showing the better results for SNR (signal to noise ratio), but it is having higher computational complexity. Therefore recently novel approach for blind channel estimation designed based on non-redundant pre-coding [1] for wireless systems in order to good performance with less SNR and less computational complexity. The approach designed in [1] is based on non redundant pre-coding was proposed in which only small fever number of sub carriers commensurate with the length of channel as well as carry the pre-coded data symbols for the purpose of blind channel estimations. With the method, the symbols transmission is done in traditional manner, hence the use of MLD (maximum likelihood detection) of data symbols as well as pre-carrier pre-coding among the antennas for the improvement in data rate. There are other more pre-coding based blind channel estimation approaches designed but suffered from the number of limitations. The joint estimation of linearly pre-coded symbols leads to the MLD method computationally hard to control. Additionally, MMSE (minimum mean square error) based estimation of symbol is highly leads to computational overhead because of need of high dimensional matrix inversion. The linearly pre-coding among all the carriers leads to very difficult task to simultaneously use the per-carrier pre-coding among the antennas for the optimization in data rate.

The existing SOS based or HOS based methods for blind channel estimations designed with different objectives. However, none of the existing techniques capable to interference signals cancellation in MIMO-OFDM [6]. Interference signals are caused by either other mobile users or fading channel in MIMO-OFDM wireless systems in blind manner. The HOS based Independent Component Analysis (ICA) method is recently designed for interference cancellation in MIMO systems, but not clearly designed and addressed for blind channel estimation.

Additionally, ISI (inter carrier interference) also having major impact of performance of blind channel estimation and spectral efficiency, which is not yet addressed. In this paper, we are designed Hybrid ICA (HICA) approach by using pulse shaping for efficient blind channel estimation in order to improve the performance of spectral efficiency and less error rates. In Sect. 2, we are discussing the different recent blind based channel estimation methods. In Sect. 3, we are presenting the system model and design of HICA blind channel estimation method. In Sect. 4, we are presenting the simulation results and comparative study. Finally the conclusion and future work is discussed in Sect. 5.

2   Related Work

This section, presents the recent methodologies introduced by different authors for blind channel estimation for MIMO-OFDM transmission systems.

In [7], author designed blind channel estimation approach using repetition index. They proposed subspace blind channel estimation method based on repetition index with similar results as compared to previous method with fever number of symbols. Author does not evaluate the computational complexity for MIMO-OFDM.

In [8], proposed the blind channel estimation method in MIMO-OFDM systems with the OSTBC (orthogonal space-time block code). They designed the new weighted covariance matrix of the data received in order to exploits the redundancies in code. They proposed this approach with aim of resolving all the non-scalar ambiguities.

In [5], another SOS based blind channel estimation method proposed. They proposed the algorithm of blind recursive for the tracking emerging time varying wireless channel in MIMO-OFDM systems which are pre-coded. With this approach subspace based tracking was designed for fast time varying wireless channels. Their approach called the data from the time as well as frequency domain as the frequency correlation of the wireless channels in order to faster the required SOSs updates.

In [6], author proposed the new blind channel estimation techniques based on subspace and SOS models for MIMO-OFDM. Their approach exploited the null space introduced by the OSTBC. This approach worked with only single receiver antenna as well. Additionally they proposed the modified proposed approach with goal of requirement of less received blocks.

In [9], another subspace based blind channel estimation method proposed for MIMO-OFDM. They designed novel signal permutation approach in this article for MIMO-OFDM. With this approach, it had high full-row-rank probability even if few OFDM symbols or low-order modulation is applied. The experimental results of this method shown that, NMSE performance is better as compared to the existing subspace methods. However, computationally having higher overhead as the interference is not handled.

In [10], new pre-coder is designed for blind channel estimation proposed for MIMO-OFDM wireless systems. With this approach, fewer number of data symbols in proportion with length of wireless channel were linearly pre-coded before the transmission. The main benefits of using this method was compatibility with improved data rate MIMO pre-coding by utilizing the less number of sub-carriers in order to introduce the signal correlation required for blind channel estimation. Interference cancellation is not addressed with this technique.

In [11], author investigated the blind channel estimation technique for MIMO-OFDM by considering the effects of channel interference and ISI. The investigation was performed with respect to BER and least square error rates.

In [12], the recent novel blind channel estimation technique designed for LTE (long term evolution) wireless networks based on advantages of wavelet transform denoising characteristics with ICA capability of blind estimation in LTE networks. They called this approach as WD-ICA. The denoising approach was designed in order to handle the blind interference cancellation. This was the first attempted approach for blind channel estimation using ICA. The ISI cancellation is not performed in this method.

3 Proposed Methodology

The system model designed for proposed HICA approach is showing in Fig. 1. Figure 1 is showing (T) transmit and (R) receive antennae. End users data is randomly generated in digital form which is first given input to pulse shaping algorithm to inter symbol interference cancellation. Before pulse shaping, first symbol mapping, IFFT/FFT and CP operations are performed at each transmitter and receiver wireless antennas. Then input symbols from the each user are modulated using any type of modulation technique like BPSK, QAM, and QPSK and transmitted over the wireless AWGN channel. Considering that there are M transmitted signals with every signal consisting of samples S. the attenuation for nth channel

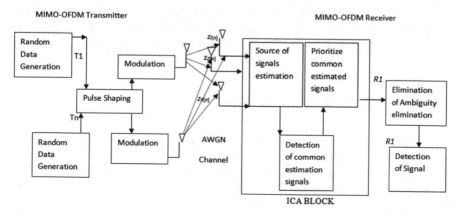

Fig. 1 Proposed HICA MIMO-OFDM system model

path is represented by *An*. Attenuation factor is complex number. The pulse shaping is introduced here to remove the ISI effects before performing the modulation in order to minimize the error rates and optimize the spectral efficiency. Then further during the transmission we designed the ICA based blind interference cancellation approach. The algorithm design and steps are elaborated in below steps.

3.1 HICA Method

The idea of the proposed method is to first apply the pulse shaping method in each users symbols in order to minimize the inter symbol interference. Pulse shape is light weight ISI cancellation approach. We used the square root raised cosine filter in up sampling domain to realize the pulse shaping filter. In this section we are just presenting the core equations those are added for efficient blind channel estimation.

In first step, let's consider $g_t(t)$ transmit side pulse shape filter for each user symbols and $g_r(t)$ is receive antenna side matched filter. The composite channel is represented as $T * R$ with matrix H (*t*). The (i_R, i_T) channel using pulse phase filtering is represented by:

$$h_{iR'}i_T(\mathrm{t}) = h_{iR'}i_{T'c}(t) * g_t(t) * g_r(t) \tag{1}$$

where $h_{iR'}i_{T'c}(\mathrm{t})$ is the (*iR, iT*) element of H (*t*).

Here the channel can be represented as the L tap FIR filters array for blind channel estimation.

In second step, for blind channel estimation is which is done by ICA block. In proposed HICA, first signals sources from receiver observation mixture are identified by using determining separation matrix represented as (*W*).

In third step, at the every iteration of HICA, the estimated signals order is different due to random initialization of HICA method. In spite of that, if there is significant information in estimated signal, then it will appear at every iteration always. The estimated common signals further prioritize based on their Higher Order Statistics (HOS) in order to select the estimated desire signals (*m*) using proposed HICA method by leaving the interference related components.

3.2 Mathematical Representation

First step is already represented above; the step 2 and 3 mathematical representation of HICA method is below:

In second step, *W* is performed using the maximizing the non-gaussianity of the observation signals principle. The non-gaussianity for random variable (*v*) containing the complex data is measured by the Kurtosis *K*[*s*] as:

$$K[v] = E\left[|v|^4\right] - 2\left(E\left[|v|^2\right]\right)^2 - E[vv]E[v^*v^*] \tag{2}$$

where $(.)^*$ is represents the complex conjugate.

The estimation of W is performed by the minimization of J (W) objective function within the unitary constraint ($WW^H = I_R$) due to negative results of kurtosis value on different modulation schemes. This objective function is based on estimated signals $\hat{s}_i$ [n] kurtosis values represented as:

$$W = \{\min_W J(W) = \sum_{j=1}^M K[\hat{s}[n]] \tag{3}$$

The objective function J(W) minimization is done by gradient computation of objective function as:

$$Jw = \frac{\partial J(W)}{\partial W} = K\left(W^H \hat{s}[n]\right)\left[E\left\{\hat{s}[n]\right)^3\right\}] \tag{4}$$

where (w) is represents the one vector from the separation matrix W.

As the objective function (Jw) optimization is in constraint of $WW^H = I_R$, object function gradient must be complemented using projecting W over the interval after each step performed by dividing the W by its norm.

In third step, the execution of HICA method number of times with various random initialization of W at each time in order to estimate the common signals for HICA executions. The two $\hat{s}_i$ [n] and $\hat{s}_i$ estimated signals for different executions are assumed different if the SAM (spectral angle mapper) among their related vectors $v_i[n]$ and $v_j[n]$ is more than the estimated threshold value ε. After that, common estimated signals are prioritized based on their 3rd and 4th HOSs as:

$$J(\hat{s}_q) = \left(\frac{1}{12}\right)[Q_q^3]^2 + \left(\frac{1}{48}\right)[Q_q^4 - 3]^2 \tag{5}$$

where, $Q_q^3 = E\left\{\hat{s}_q^3\right\} = \left(\frac{1}{T}\right)\sum_{n=1}^T (\hat{s}_q[n])^3$ is 3rd order of statistics and

$Q_q^4 = E\left\{\hat{s}_q^4\right\} = \left(\frac{1}{T}\right)\sum_{n=1}^T (\hat{s}_q[n])^4$ is 4th order of statistics of estimated signals. And q is nothing but execution index.

3.3 Ambiguity Elimination

The estimation of estimated users signals still to the permutation as well as phase rotation ambiguities due to ICA algorithm ambiguity issues. The $\hat{s}_i$ [n] is nothing similar to the original transmitted signal $s[n]$, and there is presence of ambiguity matrix A comparing with the $s[n]$. This can be represented as:

$$\breve{s}_i[n] = A \times \hat{s}_i[n] \tag{6}$$

Two indeterminacies forming the A as:

$$A = P \times D \tag{7}$$

where P is nothing but the *permutation ambiguity matrix* and (D) is *phase rotation ambiguity matrix*.

In proposed blind channel estimation method, ambiguities elimination is performed by multiplying the all estimated signals by the ambiguities elimination step of the proposed method is represent by post-multiplying the estimated signals L_F which is represented as:

$$L_F = \arg \min_{L \in G} \|s[n] - \hat{s}[n] \times A\|^2. \tag{8}$$

The final estimated signal is:

$$\hat{S}_{iF}[n] = L_F \times \hat{s}[n] \tag{9}$$

3.4 Algorithm Design

The above steps in HICA are summarized in algorithm 1 for fixed step size (μ):

Step 1: Apply the Pulse shaping on input symbols of each user in order inter symbol cancellations.
Step 2: Apply the modulation such PSK, QAM etc.
Step 3: Initialization HICA iterations in IT and set it = 0;
Step 4: Random W initialization
Step 5: Defining the objective function $J_{old} \leftarrow J(W)$.
Step 6: Gradient computation of objective function using Eq. (4).
Step 7: W updating according to negative gradient direction, $W \leftarrow W - \mu Jw$.
Step 8: W normalization according to unitary constraint, $W \leftarrow W / \|W\|$
Step 9: If $J_{old} - J(W) < \varepsilon$ (where ε is a very small threshold Value), then go back to step 4.
Step 10: Set of signals estimation $\hat{s}[n] = W^H \times Y_r$ (where, Y_r is the set of received signal)
Step 11: Form every estimated signal $\hat{s}$ [n] as vector which represented by $V_{it}[n]$
Step 12: If $it < 1$ go to step 4 else, continue.
Step 13: Find the set of the common vectors for all runs of algorithm up to itth run.

Step 14: If there is no common vectors does not appear for all the HICA executions, then go to step 4, else iteration is terminated.

Step 15: Apply Eq. (5) to prioritize the common estimated signals with J ($\hat{s}_{it}$).

Step 16: Selection of desired signals (m) with largest J ($\hat{s}_{it}$) in order perform the blind interference cancellation.

Step 17: Ambiguity Elimination using Eq. (9).

Step 18: STOP.

4 Simulation Results

The proposed blind channel estimation method is simulated and compared against the ICA based approach for BER and MSE analysis with below configuration parameters (Table 1).

4.1 MSE (Mean Square Error) Performance

Figures 2 and 3 are showing the performance of proposed approach against existing ICA method in terms of MSE for QAM and QPSK modulation techniques respectively. The proposed approach is showing the more improvement in MSE performance as compared to ICA method.

4.2 BER (Bit Rate Error) Analysis

BER is most important parameter used to evaluate the efficiency of channel estimation method. Figures 4 and 5 are showing the performance of proposed approach

Table 1 MIMO-OFDM simulation parameters

FFT Size	256
Block size	8
Sub band size	20
SNR range	0:5:50
Number of iterations	500
Channel type	Iden channel
Blind estimation method	ICA and HICA
Modulation technique	QPSK and QAM
Number of subcarriers	256
Filter type	Pulse shaping filter
Oversampling factor	4

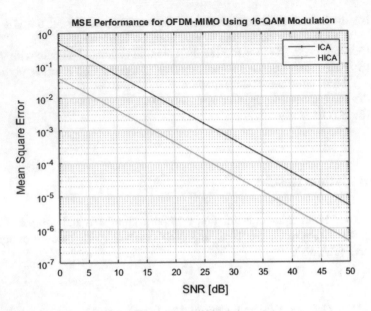

Fig. 2 MSE performance analysis using 16-QAM modulation

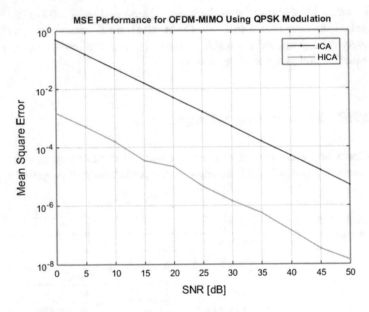

Fig. 3 MSE performance analysis using QPSK modulation

against existing ICA method in terms of BER for QAM and QPSK modulation techniques respectively. The proposed approach is showing the efficient BER performance as compared to ICA method.

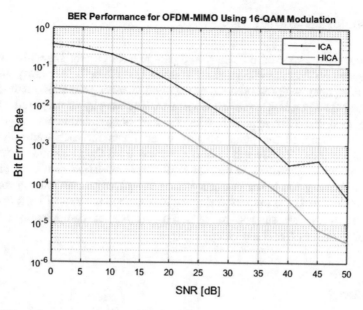

Fig. 4 BER performance analysis using 16-QAM modulation

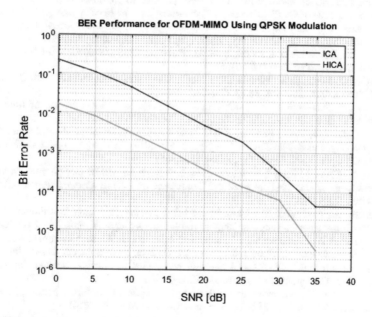

Fig. 5 BER performance analysis using QPSK modulation

5 Conclusion and Future Work

For improving the spectral efficiency and data rate for current generation wireless communication systems, efficient channel estimation methods plays very important role. In this paper, we presented the novel blind channel estimation method called HICA for MIMO-OFDM under HOS domain. The novelty of proposed approach was the modified ICA approach with ISI and blind interference cancellation. For ISI, we designed the pulse shaping based filters for each transmitting and receiving antennas. However it is also observed from the simulation results that spectral efficiency increases at cost of BER. The simulation results of proposed approach showing the better performance as compared to previous channel estimation method. HICA is outperforming the ICA method in terms of BER and MSE. For future work, we suggest to work on analysis of computation overhead, Peak-to-Average Power Ratio etc.

References

1. Tu, C.-C., Champagne, B.: Blind recursive subspace-based identification of time-varying sideband MIMO channels. IEEE Trans. Veh. Technol. **61**(2), 662–674 (2012)
2. Ngo, H., Larsson, E.G.: EVD-based channel estimation in multi cell multiuser MIMO systems with very large antenna arrays. In: Proceedings of IEEE International Conference on Acoustics, Speech, Signal Process., Kyoto, Japan, pp. 3249–3252, Mar 2012
3. Müller, R., Cottatellucci, L., Vehkaperä, M.: Blind pilot decontamination. IEEE J. Sel. Topics Signal Process. **8**(5), 773–786 (2014)
4. Shin Jr., C., Heath, R.W., Powers, E.J.: Blind channel estimation for MIMO-OFDM systems. IEEE Trans. Veh. Technol. **56**(2), 670–685 (2007)
5. Tu, C.-C., Champagne, B.: Blind recursive subspace-based identification of time-varying wideband MIMO channels. IEEE Trans. Veh. Technol. **61**(2) (2012)
6. Jiang, J.-D., Lin, T.-C., Phoong, S.-M.: New subspace-based blind channel estimation for orthogonally coded MIMO-OFDM systems. In: 2014 IEEE International Conference on Acoustic, Speech and Signal Processing (ICASSP)
7. Shao, X., Chen, J., Kuo, Y.: Blind channel estimation for MIMO-OFDM systems based on repetition index. In: 2011 International Conference on Internet Computing and Information Services. IEEE (2011)
8. Sarmadi, N., Pesavento, M.: Closed-form blind channel estimation in orthogonally coded mimo-ofdm systems: a simple strategy to resolve non-scalar ambiguities. In: 2011 IEEE 12th International Workshop on Signal Processing Advances in Wireless Communications
9. Fang, S.-H., Chen, J.-Y., Lin, J.-S., Shieh, M.-D., Hsu, J.-Y.: Subspace-based blind channel estimation for MIMOOFDM systems with new signal permutation method. In: Vehicular Technology Conference (VTC Spring). IEEE (2014)
10. Noh, S., Sung, Y., Zoltowski, M.D.: A new precoder design for blind channel estimation in MIMO-OFDM systems. IEEE Trans. Wirel. Commun. **13**(12) (2014)

11. Shirmohammadi, M., Damavandi, M.-A.: Blind channel estimation of MIMO-OFDM systems in satellite communication. In: Information and Communication Technology Convergence (ICTC). IEEE (2015)
12. Abdel-Hamid, G.M., Saad, R.S.: Blind channel estimation using wavelet denoising of independent component analysis for LTE. Indonesian J. Electr. Eng. Comput. Sci. 1(1) (2016)

Anticipated Test Design and Its Application to Evaluate and Select Embedded Libraries

Clauirton Siebra, Carla Nascimento, Leonardo Sodre,
Antônio Cavalcanti, Daniel Barros, Fernando Lima,
Fernando Cruz, Fábio Q.B. da Silva and Andre L.M. Santos

Abstract Code refactoring usually generates problems in other parts of the code that had already been validated. A solution is to use an anticipated test design methodology, where unit tests are first created to each module/class/method before their modification. Thus, developers are able to ensure the correct execution of functions after their refactoring. Our work applies this strategy to the development of a set of libraries that are used in several other projects. As developers have to modify the initial implementation of these libraries, to adapt such libraries to different needs, it is important to ensure that the libraries' functions are still properly working and verify if the efficiency of the algorithms was modified. Results show

C. Siebra (✉)
Informatics Center, Federal University of Paraiba, Joao Pessoa, Brazil
e-mail: clauirton@ci.ufpb.br

C. Nascimento · L. Sodre · A. Cavalcanti · D. Barros · F. Lima · F. Cruz
CIn/Samsung Laboratory of Research and Development, Recife, Brazil
e-mail: cmpn@cin.ufpe.br

L. Sodre
e-mail: lmas2@cin.ufpe.br

A. Cavalcanti
e-mail: avcj@cin.ufpe.br

D. Barros
e-mail: dnmlb@cin.ufpe.br

F. Lima
e-mail: flasfl@cin.ufpe.br

F. Cruz
e-mail: frlc@cin.ufpe.br

F.Q.B. da Silva · A.L.M. Santos
Centro de Informática, Universidade Federal de Pernambuco, Recife, Brazil
e-mail: fabio@cin.ufpe.br

A.L.M. Santos
e-mail: alms@cin.ufpe.br

© Springer International Publishing AG 2018
R. Lee (ed.), *Software Engineering Research, Management and Applications*,
Studies in Computational Intelligence 722, DOI 10.1007/978-3-319-61388-8_4

that this approach has increased the confidence of developers in modifying a library and generating several implementations of such library, so that the best implementation could be selected using the same set of unit tests.

Keyword Anticipated design · Test generation · Test unit · Continuous integration · Agile methods

1 Introduction

Software testing was already identified as a very labor intensive and expensive stage of the software development process, which can often account for more than 50% of total development costs [1]. While number and complexity of tests are increasing due to new resources provided by computational platforms, test centers are forced to improve their test process time. Note that as faster a specific system is evaluated and delivered to the market, as better will be its chances against other applications. This scenario configures a contradiction: the need to increase the number of tests and decrease the test time. Furthermore, this contradiction can lead to a reduction of the quality of the overall test process. Thus, there are strong benefits in reducing the cost and improving the effectiveness of the software testing process.

Test automation is the main practice discussed in the software literature as an alternative to obtain a better efficiency in software testing. In fact, there has been a rapid growth of practices in using automated software testing tools and, currently, a large number of test automation tools have been developed and have become available on the market [2]. Although automation techniques for test generation have started to be gradually adopted by the IT industry in software testing practice, there still exists a big gap between real software application systems and the practical usability of automated test generation techniques proposed by the research community. The survey presented by Rafi and colleagues [3], for example, shows that limitations of test automation are: high initial investment in automation setup, tool selection and training. Furthermore, 45% of the respondents agreed that available tools in the market offer a poor fit for their needs. Finally, it was found that 80% of the practitioners disagreed with the vision that automated testing would fully replace manual testing.

The use of automated software generation could be applied as an alternative for anticipated test design since tests cases are early and mostly generated along the specification stage. However, this type of automation is still need to leave the academy and show up its advantages when used in real problems. In addition to the limitations listed in the previous paragraph, there are several other aspects that still need to be considered. First, the maintenance of automated test cases was cited as problematic by real practitioners, mainly when the project presents an unstable set of requirements. Note that this is a common situation in the majority of real software development projects. Thus, the use of automated approaches requires the use of test cases that are highly maintainable and robust. This may be a problem because the maintenance load of automated testing is likely to increase in the future as we have already seen systems where the amount of test code exceeds the amount of production code [4]. Second, to cover initial investments, the automated approach should be easily configured and fitted to the various software development projects and ways to work. A software team will not invest in an automation process to only use it in one or few projects. In fact, the current approaches still need to better consider this issue [5]. Third, there is not a strategy that supports an incremental delivery of test automation. This lack brings problems since practitioners must directly go to a state where test automation requires high investment and then maybe provides high reward, rather than going to an initial model where test automation requires a low investment and provides a lower reward. Approaches supporting such incremental adaptation would assist to mitigate from the current high-risk reward scenario [3]. Finally, the literature supports the superiority of test automation mainly when several regression testing rounds are needed [6]. This may be the principal reason for the divergence between academic and practitioners' vision about automation. According to Rafi et al. [3], for example, while many academic sources provide evidence that test automation increases fault detection, still 58% of the practitioners do not agree with this affirmation.

Based on this discussion, our development team decided to avoid the test automation approach and use our previous experience regarding anticipated test design, which was originally proposed in [7]. The main idea is to use this approach along the development of embedded libraries that will be used in several other projects. These libraries are usually modified by developers that may not be part of the original project. Furthermore, as these libraries are usually modified in accordance with the context where they will be applied, the anticipated test design enables both the trustability of the final code and also a comparative analysis regarding the efficiency of different versions of the original library. This analysis is mainly important because it supports the selection process of the best implementation.

The remainder of this paper is organized as follows: Sect. 2 presents current approaches that use some kind of anticipated test design method and the particular features of such methods. Section 3 details the anticipated test design method defined by our development team and how it evolved from the traditional test method initially used by such team. Section 4 places the anticipated test design into the context of the evaluation processes of software libraries. Section 5 discusses our experience in a case study related to the definition of an image manipulation library. Finally, Sect. 6 concludes this work with the main remarks and research directions.

2 Approaches for Anticipated Test Design

As discussed along the introduction, the automated test generation is the most popular technique to design tests before the actual test stage. Other approaches have the same aim but, differently, do not use automation as primary method to test design.

The Built-in Test Design [8] is an example. The essential idea of the built-in test design is that component suppliers pre-place test scripts in components and set their corresponding testing-interfaces. Built-in test scripts generally contain test cases or present facilities to generate test cases which the component can use to test itself and its own methods. A component can operate in two modes: normal and maintenance. Components do not differ from other non-built-in testing enabled components while they are running in the normal mode. However, in the maintenance mode, the component user can invoke the respective component methods to execute the test, via their interfaces, evaluate the test results and output test reports [9].

Much of the work in Built-in Test Design has focused on component-based software [8–10], since reusable components have limited access and it is difficult to carry out tests in system built by externally-provided components. Our testing approach differentiates from them by providing the ability to test any arbitrary part of the system, rather than just reusable components. Note that such components are not usually produced to be modified and, consequently, their related built-in test functions are created without a special attention on the adaptability aspect.

similar approach to Built-in Test Design and derived from the hardware testing is the Design-for-Testability technique [11], where self-testing software components are produced in such way that they can autonomously evaluate themselves along, for example, integration tests. A set of associated test cases are provided with self-testing components and they are executed at the system deployment time. The successful execution of all test cases associated with a component C newly integrated in a system assures that C was correctly integrated into the system, i.e., the interactions of C with the system are correct. Similarly, the successful execution of all test cases associated with components that interact with C assures that the system integrates well with C, i.e., all interactions from the system to C are correct [12].

Design-for-Testability is again mostly focused on reusable components, where tests for software components are executed only once at deployment time since software components are not supposed to be changed during their lifetime. In addition, this approach considers integration tests as more important since unit testing must be performed during the development process. Differently, our intention is not to create test situations that will be used at deployment time, but just create facilitators in terms of testing code that accelerates the performance of the test stage. In this way, these testing codes can be implemented as both unit and integration tests, and regularly executed along different cycles.

A more general approach for anticipated test design is called Test-Driven Development (TDD), which is based on formalizing a piece of functionality as a test and implementing the functionality such that the test passes [13]. An important feature of TDD is that programmers write functional tests before the corresponding production code. The work of Muller and Hagner [14] was the first in carrying out a practical analysis of the TDD application, comparing such approach to traditional programming (design, implementation and test). The conclusions of this work turned out that TDD does not accelerate the implementation and the resulting programs are not more reliable, but test-first seems to support better program understanding. A similar result was obtained by Pancur et al. [15] who carried out an experiment with 38 senior undergraduates, showing that Test First group obtained neither higher external quality nor better code coverage. Differently, other practical studies [16, 17] support the use of TDD, so that we do not still have a clear vision about their real benefits. Thus, rather than motivate the use of this strategy, such studies show the need of more experimental analysis and the execution of replication studies.

The order of development is the main difference from our work to the TDD approach. While TDD requires the implementation of test code first than production code, we maintain the traditional order of implementation for each functional code unit. Furthermore, our tests are still executed in a well-defined test stage. Differently, TDD starts with the unit tests coding to the proposed functionalities using a unit testing framework. Afterwards, programmers write code to pass these test cases, but are not allowed to write code that does not aim to pass the tests that are already produced. Codes for the proposed functionalities are considered to be fully implemented if and only if all the existing test cases pass successfully. Note that there is a fully mixed between implementation and test stages.

The next schema (Table 1) summarizes the relation between the development process stages (specification, implementation, test and deployment) and the moments when tests are mainly created (C) and executed (E).

Table 1 Comparison between possible approaches for test design, where ASG = Automated Software Generation, BTD = Built-in Test Design, DFT = Design-For-Testability, TDD = Test-Driven Development, ATD = our approach for Anticipated Test Desigh. The symbol (C) means the main moment for test creation, while the symbol (E) means the main moment for test execution

	Specification	Implementation	Test	Deployment
ASG	(C) Tests are usually created from specification		(E) Well defined stage where code is evaluated	
BTD		(C) Provides pre place test scripts in components and sets the corresponding testing interfaces		(E) Test execution is focused on integration matters
DFT			(C) Test stage is used to create test cases for future use	(E) Test execution is focused on integration matters
TDD		(C) (E) Implementation and test stages are mixed, so that the implementation of production code is in fact a test process		
ATD		(C) Production and test code are created in pair (see Sect. 3)	(E) Well defined stage where code is evaluated	

3 Anticipated Test Design Definition

3.1 Flow of Activities

Along our previous traditional development process (Scenario A), the development team used to work in cycles. Each cycle had a set of features, which were developed according to the next flow (Fig. 1). These feature could come as result of a refactoring process (adaptation and/or modification of an existent feature), or the own system evolution (new feature). For each new feature, it was carried out an analysis of alternatives to implement it in terms of design patterns, data structures and algorithms. Then, the specification of the feature was sent to the implementation stage and, after that, to the test team, which accounted for creating test cases and evaluating the new code. If this new code passes all test set, then it is integrated to the project.

Considering this process, we have that the development time T of a feature f in this Scenario A is given by Eq. (1).

$$T_{f,A} = T_{analysis,A} + T_{implementation,A} + T_{test,A} \qquad (1)$$

In this Eq. (1), $T_{implementation,\ A}$ corresponds only to the time to implement the production code related to f, while $T_{test,\ A}$ corresponds to the sum of time to create tests cases and execute such tests. Then we have (2):

$$T_{test,A} = T_{test-implementation,A} + T_{test-execution,A} \qquad (2)$$

In Scenario B, we have a slight modification since the creation of tests is moved to the implementation stage (Fig. 2).

Considering this new process, we have that the development time T of a feature f in this scenario B is given by Eq. (3) and $T_{implementation,\ B}$ is given by Eq. (4).

$$T_{f,B} = T_{analysis,B} + T_{implementation,B} + T_{test,B} \qquad (3)$$

$$T_{implementation,B} = T_{test-implemenation,B} + T_{product-implementation,B} \qquad (4)$$

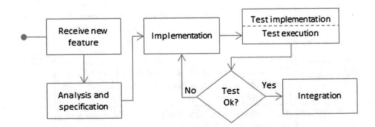

Fig. 1 Traditional flow of development (Scenario A)

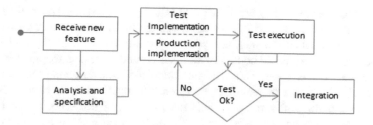

Fig. 2 Flow of development, considering an anticipated test design approach (Scenario B)

A direct implication when we change from Scenario A to Scenario B is:

$$(T_{\text{test}, B} < T_{\text{test}, A}) \text{ AND } (T_{\text{implementation}, B} > T_{\text{implementation}, A}) \quad (5)$$

Considering this equation, Scenario B will present advantages regarding Scenario A if we ensure two conditions. First, $T_{\text{test}, B}$ must be as shorter as possible than $T_{\text{test}, A}$. Second, $T_{\text{implementation}, B}$ must be as closer as possible to $T_{\text{implementation}, A}$. The first condition naturally holds, since we are leaving out the time related to test implementation from $T_{\text{test}, B}$ when compared to $T_{\text{test}, A}$. However, to ensure the second condition, we must demonstrate that $T_{\text{test-implemenation}, B}$ is significantly shorter than $T_{\text{test-implemenation}, A}$. This means:

$$T_{\text{test}-\text{implemenation}, B} < < < T_{\text{test}-\text{implementation}, A} \quad (6)$$

The theoretical rationale that supports the relation (6) in our approach says that if a developer is implementing a function, s/he is thinking about its possible problems and normal/exceptions flows. This involvement over the code elaboration gives to its developer a focused expertise, which can better lead him/her along the definition of forms of evaluation.

3.2 Definition of Miniworlds

The implementation of test code, or test methods, is only part of our test cases. The second part is related to the definition of input values to feed the tests. In our first version, each test implementation had a set of methods to create important instances in the database, considering the functionality to be tested. For example, if the method accounts for ordering a set of data by some specific field, the database should have a set of data to be ordered. So, an *initialize* function was performed to insert these data into the database. After the test, a clean-up method was performed to reset the database to the next test.

In this initial approach, the idea was to maintain the initialization methods as simple as possible. After some experiments, we noticed that some methods were still presenting errors ever when they passed through the test set. The analysis of the test cases demonstrated that the problem was the limited situations created by the input data rather than the own test codes. The improvement of the initialization methods, so that they could configure a higher number of test situations, leaded to a huge redundancy of code. This fact motivated the team to define the concept of *miniworld*, which is a metaphor to represent all input data and related expected results for a given software code. This means, a set of test scenarios that makes sense to such code and can evaluate it. Formally, given a set of test codes Ω that contains n tests $[t_1,\dots, t_n]$, where $[\alpha_1, \alpha_2,\dots, \alpha_n]$ are respectively the set of data required by $[t_1, t_2,\dots, t_n]$. A miniworld is a set of data Φ, composed by: $\alpha_1 \cup \alpha_2 \cup, \dots, \cup \alpha_n$; which tries to maximizes the coverage of Ω.

Based on the miniworld concept, the initialization is only carried out once and the test methods do not need to perform further initializations into the database. This approach significantly reduced the redundancy of test code but, on the other hand, it created an important new project thread associated with the miniworld configuration. At the moment, miniworlds are manually generated. However, the literature present some approaches to the automatic generation of databases, which present a varying rich synthetic data distribution. For example, an approach to automatically generate test data for SQL queries is described in (Suárez-Cabal et al. [18]). In this approach, the queries and database schema are used to lead the data generation. The different test situations on the queries are identified using a condition coverage criterion and they are represented with a set of constraints that the information in the database must fulfil. Other similar approaches are presented in Bruno and Chaudhuri [19] and Khalek et al. [20]. Our future works intend to analyze such techniques and their integration to our approach.

3.3 Integration with Agile Methods

The anticipated test design implements several testing principles of the Agile methods. Both approaches recognize that testing is not a separate phase, but an integral part of software development along with coding [21]. Thus, development teams use a "whole-team" approach to provide a better quality to software products. Testers on agile processes use their expertise in eliciting examples of desired behavior from customers, collaborating with the development team to turn those into executable specifications that guide coding. The anticipated test design also provides this possibility.

As supported by the anticipated test design, testing is always applied after short code modifications, so that the code evolves incrementally and interactively. This is one of the principles of Agile Methods. Thus, in contrast with other test methodologies, the anticipated test design focuses on repairing faults immediately, as suggested by the Agile methods, rather than waiting for the end of the project.

All these concepts are straightforward to support a continuous process of integration, which is the key idea of Agile testing and also explicitly supported by the anticipated test design. Thus, such approach can naturally be used by development teams that intend to apply the Agile concepts.

4 Library Evaluation and Selected Scenario

The evaluation and selection of embedded libraries to compose a software system use to be a complicated and time consuming decision making process. According to Lin et al. [22], the main reasons are:

- Difficulty in accessing applicability of libraries to the business needs of the organization due to availability of large number of libraries in the market;
- Existence of incompatibilities between various hardware and software systems;
- Lack of technical knowledge and experience to decision makers, and;
- Ongoing improvements in information technology, which are not followed by the libraries.

As the selection of inappropriate libraries can impact the quality of the application and negatively affect the success of the product in the market, there are several proposals that try to improve this activity. In general, all these proposals are based on the concept of *Multi criteria decision making* (MCDM) [23]. Taking as example the process of evaluation and selection of libraries, the goal of the MCDM involves to simultaneously consider multiple attributes to rank the available libraries and select the best one. However, this activity of library selection is usually carried out under schedule pressure and evaluators may not have time or experience to plan selection process in detail [24]. Furthermore, there is always a significant probability that libraries still need to be modified or adapted to meet the needs of the application in development [25].

Rather than looking for an appropriate library that could fit all the requirements of a given problem; a home-made or third-part set of libraries can be used as basis and be (1) adapted to different problems, (2) modified to improve aspects of efficiency, and (3) refactored due to architectural issues. In all these situations, two verifications should be carried out afterwards: correctness of the new library version and efficiency of such version when compared to its original version.

As discussed in [7], the anticipated test design is an efficient way to support these types of verifications for legacy system and this approach could also be used in this case for software libraries. Thus, the idea of our development team is to adapt or extend our libraries and use the anticipated test design approach to validate and select new definitions of the same library.

The next stages are introduced as a method to extend/adapt software libraries, given the requirements in terms of new functions or performance improvements. Such stages also show how the anticipated test design is integrated in such method:

1. Requirement definition: if we intend to select a software library, we first need to define what we expect from this library in terms of functions $(f_1, ..., f_n)$ and performance of such functions $p(f_i)$;
2. Preliminary investigation on adaptability of software libraries: if there is a Δf (lib) that represents the difference between what the library lib offers and what we expect in terms of functionalities, or/and a $\Delta p(f_i)$ that represents the difference between the real and expected performances of a function $f_i \in$ lib; then the development team needs to investigate what should be done to eliminate or attenuate Δf and Δp;
3. Short listing of extensions and/or adaptations: candidate extensions and adaptations are possible modifications that are identified to deal with Δf and Δp. Libraries can present several limitations such as essential functionalities and features that do not work with existing hardware, operating system, data management software, or network configurations. At the same time, there are several possible candidate solutions that can be used to modify the package. For example, a function for face recognition can be implemented using diverse machine learning algorithms. Thus, a list of these algorithms could be considered for detailed evaluation;
4. Definition of criteria for evaluation: the concepts of anticipated test design are very important in this stage. As unit tests of a library are created along the development of a library, or before its modification, such tests must already consider the performance criteria that are going to be used to qualify the library. Simple examples of criteria are processing time and amount of memory that is used by a process. These tests are integrated to the library and, if such library is modified, the tests can again be applied so that we have a solid way to make comparative analysis;
5. Creating new functionalities and evaluating the libraries: this stage implements the candidate functionalities according to Δf, which were specified in stage 3. Actually, this stage may work as a loop (Fig. 3), since extensions can affect other parts of the library, creating problems regarding what is provided by the signatures of the interfaces. The idea of the anticipated test design is exactly to avoid such unexpected counter-effects;
6. Adapting and evaluating functions: this stage implements the candidate changes, which were identified in stage 3. This stage can also wok as a loop (Fig. 3), where modifications are carried out and evaluated until the criteria defined in step 4 holds. In this process, for example, rating is done against each basic criterion, defined in the anticipated test design. An Aggregate score is then calculated for each library modification to support the process of selection.

Next figure (Fig. 3) illustrates how these stages are related. The first three stages (requirement definition, preliminary investigation of adaptability of software libraries and short listing of extensions and/or adaptations) are very similar to a traditional process of requirements analysis and design from software engineering.

The next two types of library modifications (functions extensions or adaptations) will be evaluated in accordance with test units already implemented, which are part

Fig. 3 Schema to apply the
anticipated test designing over
the process of adaptation of a
library to pre-defied
requirements

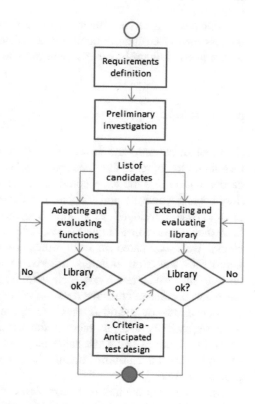

of the library. Such tests will identify both if the library is still providing their
functions in a correct way and if modifications in the internal algorithms or properties of the library have positively or negatively affected the performance of this
library. Next section discusses an application of this strategy.

5 Library Evaluation and Selection Scenario

5.1 Specification of Miniworlds to Pensieve

The case study of this paper is based on a project called *Pensieve*, whose aim is to
develop a set of libraries to analyse and select images in accordance with different
requirements of several projects related to multimedia aspects. Pensieve is divided
into a set of different libraries and each of them has a particular purpose.

Test units were introduced into each library to implement two objectives. The
first objective is to validate the level of assertiveness of each library since there is a
significant possibility that their initial implementation approaches change over the
time. Thus, the team should verify if the performance of these libraries have been
modified and quantify such modification. The second objective is to validate the

input/output of libraries' functions, using the data of pre-defined miniworlds. This process mainly ensures the correctness of functions that have been refactored or whose internal algorithms have been modified.

5.2   Method

At an initial moment, the development team did not know what could be the best techniques to compose the *Pensieve* libraries. Thus, the anticipated test design was carried out so that the team could have unit tests available and, using such tests, they could investigate and validate candidate techniques. The miniworld for this process was developed as a xml document, which had the expected results according to pre-defined inputs. Then, different approaches were evaluated and their results were compared to each other, so that we could create a ranking of approaches according different criteria. In order to do that, there were several changes in the libraries along their definitions and unit tests assisted this process since they accelerated the advances in the libraries definition because the tests could quickly validate or invalidate candidate approaches. For example, three approaches were sequentially used to modify functionalities of one of our libraries. After several evaluations using the same parameters (test units), the application of our method resulted in an improvement of about 30% in this library in terms of required processing time.

Apart the tests applied to libraries as a whole to evaluate their performances, tests were also individually applied to individual methods of libraries to ensure that the continuous changes and improvements did not affect results of other parts of the code. The test data was also created as a miniworld and the team tried to automate the generation of such miniworlds, since they could be very complex to some functions. For example, the miniworld of a functionality N was generated from a machine learning algorithm. To that end, the team used the Weka tool to specify an algorithm aimed at extracting features of a specific type of object. Then, these features could be used to classify several input objects into different clusters, creating a variety of objects to be used along evaluations.

5.3   Execution

The activities of the Pensieve were based on development cycles, which were divided into: planning phase, estimate phase, development phase and test phase. The planning (requirements definition) and estimate (preliminary investigation) phases were based on the prioritisation of functionalities, which were requested by the client. Thus, the requirements were organized according to their priorities in the development backlog, which guided the work of the development team.

The phases of development and tests were almost the same. While the team was implementing the functional code, according to the list of candidate solutions, the test units were used to validate each implementation. Functionalities were considered correct only after the application of all set of tests units. If a new function was created, a new set of test units were also created to validate this function. A miniworld with the test data was also defined or derived from an existent miniworld. An important aspect of these miniworlds was the test data for performance evaluation, which were generally defined by means of threshold values. For example, "the accuracy of functionality N must be higher than 90%". This means, such tests were the criteria to the acceptance of the implemented or modified function. Based on this approach, all functionalities of the library could be qualified according to their performances, so that the team could also have an idea about parts of the code that could be improved in case of future needs.

The creation of miniworlds tried to reuse as much as possible previous test data (input-output tuples) to avoid redundancies. This means, rather than implement new test data for each test unit, a same miniworld could be shared by different test units that require the same test data. However, the definition of this relation 1-n between test data and test units is not easy and requires a proper definition of the test interfaces.

5.4 Results

The Pensieve project has evolved from a proof of concept implementation regarding the development of libraries. Since its initial development, the team decided to use the anticipated test design so that test units were created to validate library functions along all its evolution. A qualitative analysis of the results shows that this approach resulted in gains to the project. The first gain is the higher confidence of developers, since they could try several different modifications, while ensure the code is still working. A second gain was a better understanding of the code by developers. In fact, the application of test units and the observation of its outcomes enable that developers have a better comprehension about the code even if they were not part of its development. Members of the team said that the test code uses to give important information about the functional code quality and specific features that are hard to uniquely be extracted from the functional code. This aspect positively affected the project integration, which was carried out after the complete evaluation of modified modules. Third, it was possible to monitor the libraries evolution while new adaptations were incorporated to them. The impact of the changes could also be evaluated, so that bad modifications were eliminated in an initial stage. In order, the analysis of outcomes regarding some test values could be compared to thresholds, leading the development team along the process of decision making associated with changes in the architecture and algorithms.

Table 2 compares the use of the anticipated test design in two different projects: S-Project and Pensieve Libs. S-Project is a solution to share media among devices.

Table 2 Comparative analysis of the anticipated test design application into different contexts, where SP means S-Project and PL means Pensieve Libs

Motivation to use
SP: Ensure the behaviour of existing functionalities after code refactoring
PL: Ensure the behaviour of functionalities after extensions of the library and monitor the performance of functionalities after their modification
Application of previous experiences
SP: Pioneer project, when the approach was created
PL: The experience acquired along the SP development was used. Both projects had the same technical manager, who accounted for sharing the knowledge.
Use/Adaptation of the method
SP: Development of functional code and test units with their miniworlds to evaluate a Data Access Object (DAO) layer, which was based on a legacy code.
PL: Development of functional code and test units with their miniworlds to evaluate functionalities of libraries and identify the level of performance of such functionalities.
People Involved
SP: CIn/Samsung development team
PL: Two teams physically separated, CIn/Samsung and SIDI/Samsung
Moment of inclusion of tests in the project
SP: Test units were developed along the implementation phase
PL: Test units were developed along the implementation phase

The main difference regarding these two projects is the use of tests to also evaluate and monitor the performance of several versions of the same code in the Pensieve Libs project. This further use only requires that the development of initial test units also considers tests of performance, which can evaluate the efficiency from simple methods to a complete library. After that, performance tests can be normally applied together with the remainder tests.

6 Conclusion and Research Directions

The main proposal of the anticipated test design is to implement code units that have their own test functions. Previous works show that the implementation of these functions tends to be easier since they are being implemented together to the production code that they intend to evaluate. The literature brings examples [26] of approaches that also decided by a manual creation of code units, which include their own test code. However, the practical effects of their application are not presented, so that we cannot make conclusions on their validity.

Differently, the anticipated test design was already used to assist the evolution of a legacy system and some of the advantages of its use were confirmed in our work. First, the anticipated test design enabled the definition of a test model where its parts are reusable (miniworlds). This fact contributed to reduce the time allocated to the creation of test scenarios, since it is not necessary to define data tests for each new functionality. The reuse of test data also increased the probability of capturing common failures, as reported by the project developers. Second, the anticipated test design improved the understanding of the code and a quicker assimilation of the code functionalities.

Our research directions intend to perform a quantitative analysis of this case study, showing how the use of the anticipated test design affected traditional project management metrics, such as development time, code reliability, quality and test coverage. Furthermore, we are looking for better ways to populate the miniworlds since this is a development task that is spending a significant time and the incorrect/incomplete database population may hide code problems that will compromise the final system quality [27].

Acknowledgements The authors would like to thank the support received from the SIDI/Samsung team, in particular to Helder Pinho. Professor Fabio Q. B. da Silva holds a research grant from the Brazilian National Research Council (CNPq), project #314523/2009-0.

References

1. Korel, B.: Automated software test data generation. IEEE Trans. Software Eng. **16**(8), 870–879 (1990)
2. Anand, S., Burke, E.K., Chen, T.Y, Clark, J., Cohen, M.B., Grieskamp, W., Harman, M. Harrold, M.J., McMinn, P.: An orchestrated survey of methodologies for automated software test case generation. J. Syst. Softw. **86**(8):1978–2001 (2013)
3. Rafi, D.M., Moses, K.R. K., Petersen, K., Mäntylä, M.V.: Benefits and limitations of automated software testing: Systematic literature review and practitioner survey. In: Proceedings of the 7th Int. Workshop on Automation of Software Test, pp. 36–42 (2012)
4. Berner, S., Weber, R., Keller, R.: Observations and lessons learned from automated testing. In: Proceedings of the 27th International Conference on Software Engineering, pp. 571–579 (2005)
5. Wissink, T., Amaro, C.: Successful test automation for software maintenance. In: Proceedings of the 22nd IEEE International Conference on Software Maintenance, pp. 265–266 (2006)
6. Persson, C., Yilmazturk, N.: Establishment of automated regression testing at abb: Industrial experience report on avoiding the pitfalls. In: Proceedings of the 19th IEEE International Conference on Automated Software Engineering, pp. 112–121 (2004)
7. Siebra, C., Gouveia, T., Sodre, L., Silva, F.Q.B., Santos, A.L.M.: The anticipated test design and its use in legacy code refactoring: lessons learned from a real experiment. In: 2016 International Conference on Information Technology for Organizations Development (IT4OD), Fez, pp. 1–6 (2016)
8. Beydeda, S.: Research in testing COTS components—built-in testing approaches. In: Proceedings of the 3rd ACS/IEEE International Conference on Computer Systems and Applications (2005)

9. Mao, C.: Built-in regression testing for component-based software systems. In: Proceedings of the 31st Annual International Computer Software and Applications Conference, vol. 2, pp. 723–728 (2007)

10. Wang, Y., Patel, D., King, G., Court, I., Staples, G., Ross, M., Fayad, M.: On built-in test reuse in object-oriented framework design. ACM Comput. Surv. 32(1), 7–12 (2000)

11. Binder, R.: Design for testability in object-oriented systems. Commun. ACM 37(9), 87–101 (1994)

12. Mariani, L., Pezzé, M.: A technique for verifying component-based software. In: Proceedings of the Int. Workshop on Test and Analysis of Component Based Systems. Electronic Notes in Theoretical Computer Science, vol. 116, pp. 17–30 (2005)

13. Erdogmus, H., Morisio, M., Torchiano, M.: On the effectiveness of the test-first approach to programming. IEEE Trans. Softw. Eng. 31(3), 226–237 (2005)

14. Muller, M., Hagner, O.: Experiment about test-first programming. IEEE Proc. Softw. 149(5), 131–136 (2002)

15. Pancur, M., Ciglaric, M., Trampus, M., Vidmar, T.: Towards empirical evaluation of test-driven development in a university environment. In: Proceedings of EUROCON 2003, Computer as a Tool, vol. 8, no. 2, 83–86 (2003)

16. Kaufmann, R., Janzen, D.: Implications of test-driven development: a pilot study. In: Proceedings of the 18th Annual ACM SIGPLAN Conference on Object oriented Programming, Systems, Languages, and Applications, pp. 298–299 (2003)

17. Janzen, D.: Software architecture improvement through test-driven development. In: Proceedings of the Conference on Object Oriented Programming Systems Languages and Applications, pp. 222–223 (2005)

18. Suárez-Cabal, M., De La Riva, C., Tuya, J.: Populating test databases for testing SQL queries. IEEE Lat. Am. Trans. 8(2), 164–171 (2010)

19. Bruno, N., Chaudhuri, S.: Flexible database generators. In: Proceedings of the 31st International Conference on Very large Databases, pp. 1097–1107 (2005)

20. Khalek, S., Elkarablieh, B., Laleye, Y., Khurshid, S.: Query-aware test generation using a relational constraint solver. In: Proceedings of the 23rd IEEE/ACM International Conference on Automated Software Engineering, pp. 238–247 (2008)

21. Stolberg, S.: Enabling agile testing through continuous integration. In: IEEE Agile Conference, AGILE'09, pp. 369–374 (2009)

22. Lin, H., Hsu, S., Sheen, G.: A fuzzy-based decision-making procedure for data warehouse system selection. Expert Syst. Appl. 32(3), 939–953 (2007)

23. Yoon, K., Hwang, C.: Multiple Attribute Decision-Making: An Introduction. Sage Publisher (1995)

24. Jadhava, A., Sonar, R.: Framework for evaluation and selection of the software packages: a hybrid knowledge based system approach. J. Syst. Softw. 84(8), 1394–1407 (2011)

25. Mizuno, O., Kawashima, N., Kawamoto, K.: Fault-prone module prediction approaches using identifiers in source code. Int. J. Softw. Innov. 3(1), 36–49 (2015)

26. Deveaux, D., Frison, P., Jézéquel, J.: Increase software trustability with self-testable classes in Java. In: Proceedings of the 13th Australian Conference on Software Engineering, pp. 3–11 (2001)

27. Saifan, A.A., Alsukhni, E., Alawneh, H., Sbaih, A.: Test Case Reduction Using Data Mining Technique. Int. J. Softw. Innov. 4(4), 56–70 (2016)

Improving Web Application Reliability and Testing Using Accurate Usage Models

Gity Karami and Jeff Tian

Abstract With the prevalence of the World Wide Web and its increasing size and complexity, quality assurance (QA) and testing are becoming increasingly important for web applications. Markov operational profile (Markov OP) is a good candidate for effective web quality and reliability assurance because it captures the behavior of web components and related navigation facilities to support usage based statistical testing (UBST). The accuracy of such usage models would affect the effectiveness of quality assurance and testing activities. In this paper, we examine the impact of accurate usage models on reliability, test coverage, and test efficiency. A case study is carried out to quantify this impact. We found supporting evidence that accurate Markov OP improves reliability, test coverage, and test efficiency.

Keywords Markov operational profile (OP) · Web application · Reliability · Test coverage · Test efficiency

1 Introduction

Web applications provide cross-platform universal access to web resources for the massive user population. Worldwide users rely on web applications to fulfill their needs for information processing, storage, search, and retrieval. Therefore, quality assurance (QA) for web applications is becoming increasingly important.

The concept of quality is generally associated with good user experience characterized by the absence of observable problems and satisfaction of user expecta-

G. Karami · J. Tian (✉)
Department of Computer Science and Engineering, Southern Methodist University,
Dallas, TX 75275, USA
e-mail: tian@smu.edu

G. Karami
e-mail: gkarami@smu.edu

J. Tian
School of Computer Science, Northwestern Polytechnical University,
Xi'an, Shaanxi, China

© Springer International Publishing AG 2018
R. Lee (ed.), *Software Engineering Research, Management and Applications*,
Studies in Computational Intelligence 722, DOI 10.1007/978-3-319-61388-8_5

tions [4, 6]. Quality consists of many different attributes, including reliability which is defined as the probability of failure-free operations for a specific time period or input set under a specific environment [10]. Reliability is one of the primary quality attributes for web applications [11].

Testing is a major part of quality assurance [1]. Reliability goals can be used as a objective criterion to stop testing [9]. The use of this criterion requires the testing to be performed under an environment that resembles actual usage via usage based statistical testing (UBST) to obtain realistic reliability assessment. For practical implementation of UBST, actual usage should be captured in operational profiles (OPs) [10]. OP is a usage model that quantitatively characterizes how an application will be used by its target users. Several variations of OP based on partitions, tree structures, finite state machines, and Markov OP are commonly used [13].

Markov OP is an effective and efficient way to guide UBST of a web application. Markov OP captures all components, related navigation facilities, and their usage for the web application [7]. Markov OP can help us prioritize testing effort based on usage scenarios and frequencies for individual functions and navigation patterns to improve the reliability of the web application. In addition, we can utilize the constructed Markov OP with no extra cost for traditional coverage based testing (CBT).

Accuracy of Markov OP would impact quality of the web application. A less accurate Markov OP is likely to lead to lower reliability, test coverage, and test efficiency. In this research, we present a new method to compare Markov OPs with different levels of accuracy and quantify their impact on reliability, test coverage, and test efficiency. We have applied our method in a case study to demonstrate its applicability and effectiveness.

In Sect. 2, we discuss the related work. In Sect. 3, we describe our method to quantify impact of accuracy of Markov OP on reliability, test coverage, and test efficiency. In Sect. 4, we apply our method on a case study. Finally, we discuss our conclusions in Sect. 5.

2   Related Work

In this section, we review related work in quality, reliability, quality assurance, testing techniques, and Markov OP.

2.1   Quality, Reliability, and Testing Techniques

Quality may be defined from different views, such as based on external behavior from a customer's view, or based on internal characteristic from a developer's view [6]. Quality defined based on external behavior from a customer's view can be quantified by reliability. Reliability is defined as the probability of failure-free operations for a specific time period or input set under a specific environment [10]. Two basic types of software reliability models are: input domain reliability models (IDRMs) and time domain software reliability growth models (SRGMs) [9]. IDRMs provide a snapshot of reliability. In Brown-Lipow IDRM [2], the whole input domain is partitioned into sub domains, $\{E_i, i = 1, 2, \cdots, N\}$, where each E_i represents a specific sub domain. $P(E_i)$ is the probability that input in this sub domain is used in the actual usage environment, n_i is the number of runs for this sub domain, and f_i is the number of failures observed out of n_i runs. Brown-Lipow model links OP ($P(E_i)$) to reliability, as:

$$R = 1 - \sum_{i=1}^{N} \left(\frac{f_i}{n_i} \right) P(E_i) = \sum_{i=1}^{N} R_i P(E_i)$$

$$R_i = 1 - \frac{f_i}{n_i}$$

Testing plays a central role in assuring software quality and reliability [1]. It involves executing a software and observing its results. If a failure or an observable deviation from user expectations happens, we need to locate and remove the fault or the underlying problem in the software that caused the failure. We can stop testing using reliability or coverage criterion and categorize different testing techniques accordingly into usage based statistical testing (UBST) and traditional coverage based testing (CBT). UBST views the web application from a user's perspective and focuses on the usage scenarios and associated probabilities [10]. UBST and the related reliability criterion ensure that the faults that are most likely to cause problems to users are more likely to be detected and removed, and the reliability of the software reaches certain targets before testing stops. On the other hand, CBT focuses on covering functional or implementation units and related entities [1]. CBT uses various forms of test coverage as the stopping criteria which may lead to effective fault removal. Finite state machines (FSMs) are state based models that can be used for CBT, e.g., requiring all states and state transitions be traversed [3]. Augmented FSM that include probabilistic usage information called Markov operational profile (Markov OP) can be used for UBST [13].

2.2 Markov OP Usage and Construction

Markov OP is a type of usage models for large applications involving state transitions such as web applications [7, 13]. Markov OP supports selective testing or UBST instead of complete coverage of a large web application which may be infeasible. If some components are more likely to be used, the likelihood that an underlying fault is going to be triggered through such usage is also higher. Therefore, we need to concentrate on testing highly used components. UBST using Markov OP also allows us to obtain a realistic evaluation of reliability [10]. If we have access to a Markov OP, we can also utilize it with no additional cost to do CBT on the same web application.

To construct a Markov OP for a web application, we need to identify information sources and collect data, and then identify states, transitions, input-output relations, and determine usage frequencies of individual transitions. Each state in a Markov OP can be associated with a web page or a group of web pages, and each state transition can be associated with a hyperlink or a group of hyperlinks. There are three generic methods for information gathering and OP construction, including (1) actual measurement of usage at customer installations, (2) survey of target customers, and (3) usage estimation based on expert opinion [10]. Since access logs are routinely used for existing web applications, the actual measurement of usage derived from such logs is the most effective and efficient way for obtaining usage scenarios and the corresponding Markov OP [7].

All visited web pages and hyperlinks of a web application can be collected from its access log fields, including "Requested URL" and "Referring URL". State and state transitions can be identified by assigning each web page or a group of web page to a unique state in a Markov OP, and each hyperlink or a group of hyperlinks to unique state transitions. Transition probabilities can also be calculated using these fields. Therefore, Markov OPs can be constructed based on the access log following the method developed in [7].

2.3 Maintaining Accuracy of Markov OP

Markov OPs constructed by different people using different methods and information sources may have different levels of accuracy. Accuracy of Markov OPs may also deteriorate over maintenance and evolution. The initial Markov OP constructed before maintenance and evolution may not accurately reflect the actual usage of the updated web application. At this point, the updated web application has not been deployed yet, so that its actual usage data could not be collected to construct a new Markov OP. If we utilize the initial Markov OP to do UBST or CBT of the updated web application, it may negatively affect its reliability, test coverage, and test efficiency.

A new method was developed in previous research to maintain accuracy of Markov OP over maintenance and evolution [8]. Since the user behavior reflected in the initial Markov OP is not expected to change drastically after maintenance activities, the initial Markov OP is utilized as a starting point in this method for the updated Markov OP. On the other hand, activity diagrams commonly used in software development describe the application in terms of activities [5]. Such models share some common characteristic with Markov OP. In this method, the initial Markov OP is compared with activity diagrams to examine effects of maintenance and evolution on the usage of the updated web application. The results from this comparative analysis is utilized to update the structure and transition probabilities of the initial Markov OP. This method to maintain the accuracy of Markov OP was validated by a case study to show that the updated Markov OP was found more accurate than the initial Markov OP.

3   Impacts of Accuracy of Markov OP

In this section, we first quantify Markov OP accuracy. Then, we develop a method to quantify the impact of accuracy of Markov OP on test coverage, test efficiency, and reliability.

3.1   Quantifying Markov OP Accuracy

Markov OP has three basic elements: states, transitions, and transition probabilities. In this section, we introduce the following notations to represent a Markov OP and its elements:

- $< S, T, P >$: a Markov OP.
- $S = \{s_i | i = 1, 2, \cdots, n\}$, s_i: a state.
- $T = \{t_{ij} | i, j = 1, 2, \cdots, n\}$, t_{ij}: a transition from s_i to state s_j.
- $P = \{p_{ij} | i, j = 1, 2, \cdots, n\}$, p_{ij}: the transition probability from state s_i to state s_j.
- $|S|$: cardinality of set S.

Markov OPs may have different levels of accuracy, as they are constructed by different people using different methods and information sources at different times. We quantify accuracy of a given Markov OP by comparing it with a reference Markov OP which is assumed to be 100% correct. We consider the following scenarios:

1. Common states or transitions: When subset S_c or subset T_c is present in both the reference Markov OP and the given Markov OP, we consider S_c or T_c as a common subset of states or transitions. In Fig. 1, subset S_c is the common subset of states between the reference Markov OP and the other Markov OPs.
2. Missing states or transitions: When subset S_m or subset T_m is absent in the given Markov OP while it is present in the reference Markov OP, we consider S_m or T_m as a subset of missing states or transitions in the given Markov OP. For example, subset S_m is absent in the given Markov OP with missing states in Fig. 1b, while it is present in the reference Markov OP in Fig. 1a.
3. Extra states or transitions: When subset S_e or subset T_e is present in the given Markov OP while it is absent in the reference Markov OP, we consider subset S_e or T_e as an extra subset of states or transitions in the given Markov OP. For example, subset S_e is present in the given Markov OP with extra states in Fig. 1c, while it is absent in the reference Markov OP in Fig. 1a.
4. Incorrect states or transitions: We treat the subset of incorrect states as a combination of a subset of missing states and a subset of extra states. We also treat the subset of incorrect transitions as a combination of a subset of missing transitions and a subset of extra transitions. Figure 1d shows a Markov OP with incorrect states. In this Markov OP, subset S_m is the corresponding subset of missing states and subset S_e is the corresponding subset of extra states.

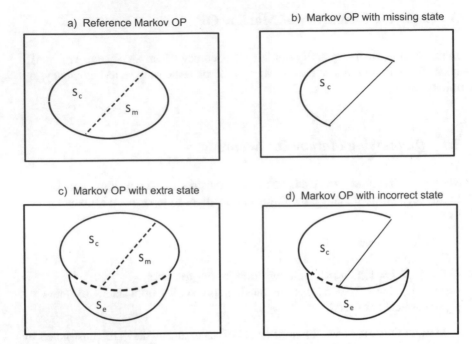

Fig. 1 Markov OPs with lower level of accuracy than reference Markov OP

5. Incorrect transition probabilities: Subset T_x in the given Markov OP has incorrect transition probabilities, if its corresponding subset in the reference Markov OP has different transition probabilities.

By quantifying missing states and transitions, extra states and transitions, states with incorrect probabilities in the given Markov OP, we can assess the accuracy of the given Markov OP.

3.2 Impact on Test Coverage

As we discussed in the previous section, if we have already constructed a Markov OP for a web application, we can utilize it to perform traditional CBT with no extra cost. By definition, the reference Markov OP can achieve 100% test coverage. If we utilize a given Markov OP with a lower level of accuracy to perform CBT for a web application, we may reach a different level of test coverage than that achieved by using the reference Markov OP. In this section, we examine the impact of accuracy of Markov OP on test coverage for web applications.

The given Markov OP may have missing states or transitions, so we can not perform CBT on the corresponding missing components or links. For example, if subset

S_m or subset T_m is a subset of missing states or transitions in the given Markov OP, we can not perform CBT on the corresponding components or links in the web application. Therefore, there is a direct link between missing states or transitions in the given Markov OP and reduced coverage. We calculate state coverage V_s, transition coverage V_t, and overall coverage V for the given Markov OP using the following equations:

$$V_s = 1 - \frac{|S_m|}{|S|} = \frac{|S| - |S_m|}{|S|} = \frac{|S_c|}{|S|}$$

$$V_t = 1 - \frac{|T_m|}{|T|} = \frac{|T| - |T_m|}{|T|} = \frac{|T_c|}{|T|}$$

$$V = \frac{(|S| + |T|) - (|S_m| + |T_m|)}{|S| + |T|} = \frac{|S_c| + |T_c|}{|S| + |T|}$$

By comparing test coverage for the given Markov OP and the reference Markov OP, we can conclude that the given Markov OP with a lower level of accuracy leads us to a reduced coverage to the levels specified in the equations above, down from 100% when the reference Markov OP is used.

3.3 Impact on Test Efficiency

The given Markov OP may have extra states or transitions, so we may end up wasting time to perform CBT on extra components or links not present in the reference Markov OP. For example, if subset S_e or subset T_e is a subset of extra states or transitions in the given Markov OP, we waste time performing CBT on the corresponding components or links in the web application. Therefore, there is a direct link between extra states or transitions in the given Markov OP with reduced test efficiency and increased cost. We calculate the relative state waste W_s, transition waste W_t, and overall waste W for the given Markov OP using the following equations:

$$W_s = \frac{|S_e|}{|S|}$$

$$W_t = \frac{|T_e|}{|T|}$$

$$W = \frac{|S_e| + |T_e|}{|S| + |T|}$$

The reference Markov OP has no waste. By comparing test efficiency for the given Markov OP and the reference Markov OP, we can conclude that the given Markov OP with lower level of accuracy leads us to a reduced efficiency as characterized by the amount of relative waste given in the above equations.

3.4   Impact on Reliability

As we discussed in Sect. 2, Markov OP can be used to perform UBST and to obtain a realistic evaluation of reliability of a web application. If we utilize a reference Markov OP and a given Markov OP with a lower level of accuracy, we may reach different levels of reliability for the web application. In this section, we discuss the impact of accuracy of Markov OP on reliability of the web application. Before examining their impact on reliability, we introduce the following terms and notations:

- R^0: Reliability of a web application before testing.
- R^x: Reliability of a web application after testing and removing detected faults based on Markov OP-x. In particular, R^r is the reliability of the web application after testing based on the reference Markov OP, and R^g is reliability of the web application after testing based on a given Markov OP.

The given Markov OP may include common states or transition, missing states or transition, extra states or transitions, incorrect states or transition, and incorrect probabilities. Common states, extra states, incorrect states, and missing states have primary impact on the reliability, while the others have secondary impact on the reliability. In the following, we discuss the primary impact in details. We plan to address the secondary impact in the future.

- Common states: If subset S_c is present in both the reference Markov OP and the given Markov OP, we can test components in the web applications corresponding to subset S_c using the corresponding Markov OP. If we utilize a Markov OP to test the web application, we may detect and remove faults from the components corresponding to subset S_c. If we remove the detected faults, R_c^r and R_c^g would be higher than R_c^0. In addition, $R_c^r \approx R_c^g$, as we are not considering the secondary impact on reliability by differences in probability distributions associated with the subset S_c for different Markov OPs. Therefore, we can take advantage of reliability growth resulted from UBST using either the reference Markov OP or the given Markov OP.
- Missing states: If subset S_m is not present in the given Markov OP while it is present in the reference Markov OP, we can not test components in the web applications corresponding to subset S_m using the given Markov OP and the number of underlying faults remain unchanged. Since we do not remove the faults from web pages corresponding to the subset S_m, R_m^g would remain the same as R_m^0. Therefore, we can not get full benefit of reliability growth resulted from UBST using the given Markov OP with missing states. On the other hand, the reference Markov OP

doesn't have any missing state. If we utilize the reference Markov OP to test the web application, we may detect and remove faults which have not been detected by the given Markov OP. If we remove the detected faults, R_m^r would be higher than R_m^0 and R_m^g, where $R_m^g = R_m^0$ as stated above. Therefore, we can take advantage of reliability growth associated with subset S_m resulted from UBST using the reference Markov OP, but not from UBST using the given Markov OP.

- Extra states: If subset S_e is present in the given Markov OP while it is not present in the reference Markov OP, we may observe some failures resulted from components in the web applications corresponding to subset S_e using the given Markov OP. However, the probability that the components in the web applications are under actual usage is zero, because S_e is not present in the reference Markov OP. So removing the underlying faults associated with subset S_e is a waste of time and doesn't improve the reliability of the web application. Therefore, we can not get the benefit of reliability growth resulted from UBST using the given Markov OP on these extra states. On the other hand, the reference Markov OP doesn't have any extra state or transition, resulting in no wasted testing effort that has no impact on reliability.

After applying any Markov OP to perform UBST, the reliability is then evaluated under the actual usage environment captured by the reference Markov OP. We apply Brown-Lipow model to quantify reliability of the web application in 3 steps:

1. We first need to partition the web application into sub domains and calculate reliability for each sub domain. We partition the web application into components corresponding to subset S_c, subset S_m, and subset S_e. R^0 and R^x can be calculated for these subsets based on pre and post testing data.
2. We need to calculate $P(S_c)$, $P(S_m)$, and $P(S_e)$. To avoid the theoretical difficulties of of non-stationary stochastic processes [12], we estimate probability of each subset based on frequency of states in each subset from actual usage of the web application as captured by the reference Markov OP. Although we may observe some failures resulted from components in the web applications corresponding to subset S_e, frequency of subset S_e in the actual usage as captured by the reference Markov OP is zero, so $P(S_e)=0$. Therefore, we only need to estimate $P(S_c)$ and $P(S_m)$ here.
3. We can estimate reliability of the whole web application for each Markov OPs by applying Brown-Lipow model to corresponding pre and post testing reliability for each sub domain.

We use the following equations to quantify reliability of the web application after testing based on the reference Markov OP and the given Markov OP:

$$R^g = R_c^g P(S_c) + R_m^0 P(S_m) + R_e^g P(S_e)$$

$$R^r = R_c^r P(S_c) + R_m^r P(S_m)$$

Since $P(S_e) = 0$, R^g is reduced to :

$$R^g = R_c^g P(S_c) + R_m^0 P(S_m)$$

In addition, we can not get the benefit of reliability growth resulted from UBST using the given Markov OP with missing states as captured by R_m^0 in the above equation. As stated in the reliability analysis earlier, $R_m^r > R_m^0$ and $R_c^r \approx R_c^g$. Therefore, we can conclude that post reliability using the reference Markov OP would most likely be higher than post reliability based on the given Markov OP with a lower level of accuracy.

4   Results from a Case Study

In this section, we first provide a case study and discuss the experimental setup. Then, we apply our method on the case study to examine the impact of accuracy of Markov OP on test coverage, test efficiency, and reliability.

4.1   Case Study

As we mentioned in Sect. 2, accuracy of Markov OP deteriorates after maintenance and evolution. In our previous research, we developed a new method to update an initial Markov OP constructed before maintenance by analyzing its differences with activity diagrams [8].

We used a student payments (SP) web application as a case study. SP helped international students make payments to register in different exams. An initial Markov OP was constructed based on actual usage of the web application before maintenance. Then, the initial Markov OP was updated using our method. Finally, a new Markov OP was constructed based on actual usage of the updated web application after its deployment. By comparing the Markov OPs, we validated that the updated Markov OP is more accurate than the initial Markov OP.

Figure 2 shows the high level initial Markov OP constructed based on actual usage of SP web application before maintenance activities. Figure 3 shows the high level updated Markov OP constructed using the method described in [8]. Figure 4 shows the high level new Markov OP constructed based on actual usage of SP web application after maintenance activities. In this paper, we utilize tabular presentation for lower level Markov OPs, as it is easier for comparing them for missing, extra, and incorrect states and transitions. Tables 1, 2, and 3 show elements of the initial Markov OP, the updated Markov OP, and the new Markov OP for the GRESubset.

Since the new Markov OP is constructed based on the actual usage of the updated web application, we consider it the reference Markov OP. We applied our method on the updated Markov OP by comparing it to the reference Markov OP to examine

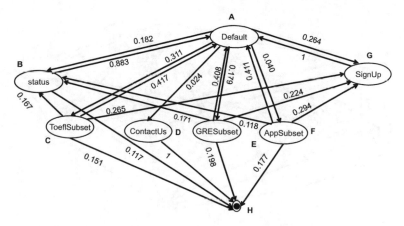

Fig. 2 Initial Markov OP-L for SP as a whole

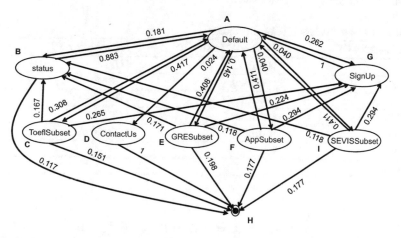

Fig. 3 Updated Markov OP for SP as a whole

the impact of its accuracy on test coverage, test efficiency and reliability. We also applied our method on the initial Markov OP. In the following, we provide results of applying our method on these Markov OPs.

4.2 Results for the Updated Markov OP

We first compared the updated Markov OP with the reference Markov OP to assess its accuracy. We found that the updated Markov OP doesn't have any missing, extra, and incorrect state or transition. Therefore, the updated Markov OP has the same level of accuracy in overall structure as the reference Markov OP. The list of common, missing, and extra states or transitions are:

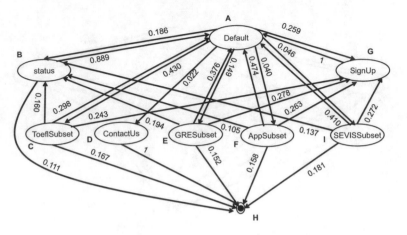

Fig. 4 New Markov OP for SP as a whole

Table 1 States in Markov OPs for GRESubset

Initial Markov OP	Updated Markov OP	New Markov OP
GRE	GRE	GRE
GREReg	GREReg	GREReg
GRERegV	GRERegV	GRERegV
GRERep	GRERep	GRERep
GRERepV	GRERepV	GRERepV
GRERes		
GREResc		

$$S_c = S, S_e = \emptyset, S_m = \emptyset, T_c = T, T_e = \emptyset, T_m = \emptyset$$

Then, we quantified state coverage, transition coverage, overall coverage for the updated Markov OP using the formulas provided in the previous section. We found that state coverage, transition coverage, and overall coverage for the updated Markov OP are:

$$V_s = \frac{|S_c|}{|S|} = 1$$

$$V_t = \frac{|T_c|}{|T|} = 1$$

$$V = \frac{|S_c| + |T_c|}{|S| + |T|} = 1$$

Therefore, we reached the same level of test coverage based on the updated Markov OP or the reference Markov OP.

Table 2 Transitions in Markov OPs for GRESubset

Initial Markov OP	Updated Markov OP	New Markov OP
$T_{GRE,GREReg}$	$T_{GRE,GREReg}$	$T_{GRE,GREReg}$
$T_{GREReg,GRERegV}$	$T_{GREReg,GRERegV}$	$T_{GREReg,GRERegV}$
$T_{GREReg,Signup}$	$T_{GREReg,Signup}$	$T_{GREReg,Signup}$
$T_{GRERegV,Default}$	$T_{GRERegV,Default}$	$T_{GRERegV,Default}$
$T_{GRERegV,Status}$	$T_{GRERegV,Status}$	$T_{GRERegV,Status}$
$T_{GRERegV,ENDState}$	$T_{GRERegV,ENDState}$	$T_{GRERegV,ENDState}$
$T_{GRE,GRERep}$	$T_{GRE,GRERep}$	$T_{GRE,GRERep}$
$T_{GREReg,GRERepV}$	$T_{GRERep,GRERepV}$	$T_{GRERep,GRERepV}$
$T_{GRERep,Signup}$	$T_{GRERep,Signup}$	$T_{GREReg,Signup}$
$T_{GRERepV,Default}$	$T_{GRERepV,Default}$	$T_{GRERepV,Default}$
$T_{GRERepV,Status}$	$T_{GRERepV,Status}$	$T_{GRERepV,Status}$
$T_{GRERepV,ENDState}$	$T_{GRERepV,ENDState}$	$T_{GRERepV,ENDState}$
$T_{GRE,SignUp}$	$T_{GRE,SignUp}$	$T_{GRE,SignUp}$
$T_{GRE,GRERes}$		
$T_{GRERes,Default}$		
$T_{GRERes,ENDState}$		
$T_{GRE,GREResc}$		
$T_{GREResc,Default}$		
$T_{GREResc,ENDState}$		

We also quantified the relative state waste, transition waste, and overall waste of the web application for the updated Markov OP using the formulas provided in the previous section. We found that the relative state waste, transition waste, and overall waste for the updated Markov OP are:

$$W_s = \frac{|S_e|}{|S|} = 0$$

$$W_t = \frac{|T_e|}{|T|} = 0$$

$$W = \frac{|S_e| + |T_e|}{|S| + |T|} = 0$$

Therefore, we reached the same level of the test efficiency based on the updated Markov OP or the reference Markov OP.

Finally, we quantified reliability of the web application for the updated Markov OP using the formulas provided in the previous section. As for reliability we only consider the primary impact due to missing, extra, and incorrect states, the estimated reliability resulting from the updated Markov OP would also be the same as that from the reference Markov OP, as follows:

Table 3 Transition Probabilities in Markov OPs for GRESubset

Initial Markov OP	Updated Markov OP	New Markov OP
$P_{GRE,GREReg} = 0.578$	$P_{GRE,GREReg} = 0.709$	$P_{GRE,GREReg} = 0.750$
$P_{GREReg,GRERegV} = 0.796$	$P_{GREReg,GRERegV} = 0.796$	$P_{GREReg,GRERegV} = 0.760$
$P_{GREReg,Signup} = 0.204$	$P_{GREReg,Signup} = 0.204$	$P_{GREReg,Signup} = 0.240$
$P_{GRERegV,Default} = 0.486$	$P_{GRERegV,Default} = 0.486$	$P_{GRERegV,Default} = 0.512$
$P_{GRERegV,Status} = 0.286$	$T_{GRERegV,Status} = 0.286$	$P_{GRERegV,Status} = 0.269$
$P_{GRERegV,ENDState} = 0.228$	$P_{GRERegV,ENDState} = 0.228$	$P_{GRERegV,ENDState} = 0.219$
$P_{GRE,GRERep} = 0.171$	$P_{GRE,GRERep} = 0.210$	$P_{GRE,GRERep} = 0.194$
$P_{GREReg,GRERepV} = 0.770$	$P_{GRERep,GRERepV} = 0.770$	$P_{GRERep,GRERepV} = 0.786$
$P_{GRERep,Signup} = 0.230$	$P_{GRERep,Signup} = 0.230$	$P_{GREReg,Signup} = 0.214$
$P_{GRERepV,Default} = 0.500$	$P_{GRERepV,Default} = 0.500$	$P_{GRERepV,Default} = 0.546$
$P_{GRERepV,Status} = 0.300$	$P_{GRERepV,Status} = 0.300$	$P_{GRERepV,Status} = 0.273$
$P_{GRERepV,ENDState} = 0.200$	$P_{GRERepV,ENDState} = 0.200$	$P_{GRERepV,ENDState} = 0.181$
$P_{GRE,SignUp} = 0.066$	$P_{GRE,SignUp} = 0.081$	$P_{GRE,SignUp} = 0.056$
$P_{GRE,GRERes} = 0.079$		
$P_{GRERes,Default} = 0.667$		
$P_{GRERes,ENDState} = 0.333$		
$P_{GRE,GREResc} = 0.106$		
$P_{GREResc,Default} = 0.625$		
$P_{GREResc,ENDState} = 0.375$		

$$R^r = R^u = R_c^r P(S_c) + R_m^r P(S_m) = R_c^r$$

Because $P(S_m) = 0$, $P(S_c) = 1$, and $S_c = S$ in this case.

4.3 Results for the Initial Markov OP

We first compared the initial Markov OP with the reference Markov OP to assess accuracy of the initial Markov OP. We found common, missing, and extra states and transitions between the reference Markov OP and the initial Markov OP, as characterized below:

$S_c = \{Default, SignUp, status, ToeflSubset, ContactUS, AppSubset, GRESubset, GRE, GREReg, GRERegV, GRERep, GRERepV\}$

$T_c = \{T_{Default,SignUp}, T_{Default,status}, T_{GRE,Signup}, T_{Default,ToeflSubset}, T_{Default,ContactUS}, T_{Default,AppSubset}, T_{Default,GRESubset}, T_{SignUp,Default}, T_{ToeflSubset,SignUp}, T_{status,ENDState}, T_{ToeflSubset,Default}, T_{status,Default}, T_{ToeflSubset,status}, T_{ToeflSubset,ENDState}, T_{GRERepV,Status},$

$T_{GRESubset,SignUp}, T_{GRESubset,status}, T_{GRERegV,Status}, T_{AppSubset,Default}, T_{AppSubset,SignUp},$
$T_{AppSubset,status}, T_{AppSubset,ENDState}, T_{GRE,GREReg}, T_{GREReg,GRERegV}, T_{GRERegV,Default},$
$T_{GREReg,Signup}, T_{GRESubset,ENDState}, T_{GRERegV,ENDState}, T_{GRE,GRERep}, T_{GREReg,GRERepV},$
$T_{GRERep,Signup}, T_{GRERepV,Default}, T_{GRESubset,Default}, T_{GRERepV,ENDState}, T_{ContactUs,ENDState}\}$

$S_m = \{SEVISSubset\}$

$T_m = \{T_{SEVISSubset,Default}, T_{Default,SEVISSubset}, T_{SEVISSubset,status}, T_{SEVISSubset,SignUp},$
$T_{SEVISSubset,ENDState}\}$

$S_e = \{GRERes, GREResc\}$

$T_e = \{T_{GRE,GRERes}, T_{GRE,GREResc}, T_{GRERes,Default}, T_{GRERes,ENDState}, T_{GREResc,Default},$
$T_{GREResc,ENDState}\}$

Then, we quantified state coverage, transition coverage, and overall coverage for the initial Markov OP using the formulas derived in the previous section. The state coverage, transition coverage, and overall coverage for the reference Markov OP are all 1. However, the initial Markov OP leads us to a lower test coverage as follows:

$$V_s = \frac{|S_c|}{|S|} = \frac{12}{13} = 0.92$$

$$V_t = \frac{|T_c|}{|T|} = \frac{35}{40} = 0.87$$

$$V = \frac{|S_c| + |T_c|}{|S| + |T|} = \frac{12 + 35}{13 + 40} = 0.89$$

We also quantified the relative state waste, transition waste, and overall waste of the web application for the initial Markov OP using the formulas provided in the previous section. The relative state waste, transition waste, and overall waste for the reference Markov OP are all 0. However, the initial Markov OP leads us to a lower test efficiency as follows:

$$\frac{|S_e|}{|S|} = \frac{2}{13} = 0.15$$

$$\frac{|T_e|}{|T|} = \frac{6}{40} = 0.15$$

$$\frac{|S_e| + |T_e|}{|S| + |T|} = \frac{2 + 6}{13 + 40} = 0.15$$

By analyzing the results, we can conclude the initial Markov OP with lower level of accuracy leads us to a reduced coverage and efficiency.

After quantifying test coverage and test efficiency, we quantified reliability of the web application based on the initial Markov OP using the formulas provided in the previous section. For reliability we only consider the primary impact due to missing, extra, and incorrect states. Based on actual usage data after the web maintenance activities, we calculated that

$$P(S_c) = 0.95, P(S_m) = 0.05, \text{ and } P(S_e) = 0$$

The estimated reliability resulting from the initial Markov OP and the reference Markov OP would be as follows:

$$R^i = R^i_c * 0.95 + R^0_m * 0.05$$

$$R^r = R^r_c * 0.95 + R^r_m * 0.05$$

We found that $R^r > R^i$ because of the following reasons:

- The reference Markov OP leads us to remove the faults corresponding to subset S_m, but we can not take advantage of reliability growth of subset S_m based on the initial Markov OP. Therefore, $R^r_m > R^0_m$.
- The reference Markov OP and the initial Markov OP lead us to remove the faults corresponding to subset S_c, so $R^r_c \approx R^i_c$.

Therefore, we can conclude the initial Markov OP with lower level of accuracy leads us to a lower level of reliability than the reference Markov OP.

4.4   Summary of Case Study Results

Table 4 shows state coverage, transition coverage, and overall coverage for the initial Markov OP, the updated Markov OP, and the reference Markov OP. Table 5 shows the relative state waste, transition waste, and overall waste for the initial Markov OP, the updated Markov OP, and the reference Markov OP. Table 6 shows the reliability of the web application based on the initial Markov OP, the updated Markov OP, and the reference Markov OP.

To summarize, after applying our method on the reference Markov OP and the updated Markov OP, we found that the reference Markov OP and the updated Markov OP lead us to the same level of test coverage, test efficiency, and reliability as shown in Tables 4, 5, and 6. After applying our method on the reference Markov OP and the initial Markov OP, we found that the initial Markov OP with lower levels of accuracy leads us to a lower level of test coverage, test efficiency, and reliability than the reference Markov OP also shown in these tables.

As characterized in this case study as well as in [8], the initial Markov OP is less accurate than the updated Markov OP, and the updated Markov OP has no missing or

Table 4 Test coverage for Markov OPs with different levels of accuracy

	State coverage	Transition coverage	Overall coverage
Reference Markov OP	1	1	1
Updated Markov OP	1	1	1
Initial Markov OP	0.92	0.87	0.89

Table 5 Waste for Markov OPs with different levels of accuracy

	State waste	Transition waste	Overall waste
Reference Markov OP	0	0	0
Updated Markov OP	0	0	0
Initial Markov OP	0.15	0.15	0.15

Table 6 Reliability based on Markov OPs with different levels of accuracy

	Reliability
Reference Markov OP	$R^r = R^r_c * 0.95 + R^r_m * 0.05$
Updated Markov OP	$R^u = R^r$
Initial Markov OP	$R^i = R^i_c * 0.95 + R^0_m * 0.05, R^i < R^r$

extra states or transitions. Therefore, this case study demonstrated that less accurate Markov OPs lead to reduced test coverage, test efficiency, and reliability, which can be quantified by the equations we derived in this paper.

5 Conclusion

Markov OP of a web application can be used to perform usage based statistical testing (UBST) and to assess web application reliability. In addition, if we have access to a Markov OP, we can utilize it for traditional coverage based testing (CBT). A less accurate Markov OP may lead us to lower test coverage, lower test efficiency, and lower reliability. In this paper, we developed a new method to quantify the impact of accuracy of Markov OP on test coverage, test efficiency, and reliability.

In our method, we assessed the accuracy of a given Markov OP by comparing it with the reference Markov OP which is assumed to be 100% accurate. We identified subsets of missing, extra, and incorrect states and transitions. Then, we calculated test coverage and test efficiency for each Markov OP using formulas we derived from the comparative analysis. We quantified primary impact of accuracy of Markov OP on reliability due to missing, extra, and incorrect states. We applied Brown-Lipow model to quantify the overall reliability of the web application. We applied our method on a case study to demonstrate that a Markov OP with a lower level of accuracy leads us to lower test coverage, lower test efficiency, and lower reliability.

In this paper, we examined impact of accuracy of Markov OP on test coverage and test efficiency due to missing, extra, and incorrect states or transitions. We also examined primary impact of accuracy of Markov OP on reliability due to missing, extra, and incorrect states. As a follow up to this study, we plan to address the secondary impact of accuracy of Markov OP due to other differences between the given Markov OP and the reference Markov OP in the future. We also plan to address impact of accuracy of Markov OP on usability and customer satisfaction.

Markov OP not only help us perform UBST and CBT, but can also be used to understand user behavior, and fine-tune system performance and usability. Therefore, accuracy of Markov OP may affect test coverage, test efficiency, reliability, usability, customer satisfaction, and communication. In conclusion, based on the analysis and the case study in this paper, we can help improve web application reliability, test coverage, and test efficiency by constructing and maintaining accurate Markov OPs. Such accurate Markov OPs can also contribute to an overall improvement to web application quality and user satisfaction.

Acknowledgements This work is supported in part by National Science Foundation (NSF) Grant #1126747 and NSF Net-Centric I/UCRC.

References

1. Beizer, B.: Software Testing Techniques. Van Nostrand Rinhold (1983)
2. Brown, J.R., Lipow, M.: Testing for software reliability. In: Proceedings of the International Conference on Reliable Software, pp. 518–527 (1975)
3. Chow, T.S.: Testing software design modeled by finite-state machines. IEEE Trans. Softw. Eng. **4**(3), 178–187 (1978)
4. Denning, P.J.: What Is software quality? Communications of the ACM **35**(1), 13–15 (1992)
5. Eshuis, R.: Symbolic model checking of UML activity diagrams. IEEE Trans. Softw. Eng. **15**(1), 1–38 (2006)
6. ISO/IEC 25010 System and Software Engineering—Systems and Software Quality Requirements and Evaluation (SQuaRE)—System and Software Quality Models, ISO (2011)
7. Kallepalli, C., Tian, J.: Measuring and modeling usage and reliability for statistical web testing. IEEE Trans. on Softw. Eng. **27**(11), 1023–1036 (2001)
8. Karami, G., Tian, J.: Maintaining Accurate Web Usage Models Using Updates from Activity Diagrams, Submitted to Information and Software Technology (2017)
9. Lyu, M.R.: Software Reliability Engineering. IEEE Computer Society Press and Mcgraw-Hill (1996)
10. Musa, J.D.: Software Reliability Engineering. McGraw-Hill (1998)
11. Offutt, J.: Quality attributes of web software applications. IEEE Softw. **19**(2), 25–32 (2003)
12. Taylor, H.M., Karlin, S.: An Introduction to Stochastic Modeling, 3rd edn. Academic Press (1998)
13. Whittaker, J.A., Thomason, M.G.: A markov chain model for statistical software testing. IEEE Trans. Softw. Eng. **42**(10), 812–824 (1994)

C-PLAD-SM: Extending Component Requirements with Use Cases and State Machines

Kevin A. Gary and M.B. Blake

Abstract Classic approaches to component specification derived from component requirements emphasize identifying external interfaces and behaviors. The C-PLAD requirements model provided a unifying framework for combining domain requirements and application requirements through an iterative refinement process. C-PLAD repackaged UML features and Unified Process techniques into an iterative process. In our continuing work, we found another layer was required—the inclusion of state machines to drive the architectural specifications beyond component interfaces and into component states in order to provide guarantees in our domains of interest, namely safety-critical applications. In this paper we describe an extension to the C-PLAD approach, dubbed C-PLAD-SM, which addresses the gaps in our earlier work.

Keywords Component · State machine · Architecture

1 Introduction

C-PLAD is a development process based on the Unified Process (RUP) [13]. The major innovation is that this process jointly supports domain engineering and application engineering within one analysis, design and development process. C-PLAD is divided into six high-level phases (see Fig. 1). These phases are Specification, Requirements, High-Level Use Cases, Component-Level Use Cases, Software Design and Development, and Testing. The software design and development phase and the testing phase are iterative phases. In the Specification Phase,

K.A. Gary (✉)
The School of Computing Informatics, and Decision Systems Engineering,
The Ira A. Fulton Schools of Engineering, Arizona State University, Mesa, AZ 85281, USA
e-mail: kgary@asu.edu

M.B. Blake
College of Computing & Informatics, Drexel University, Philadelphia, PA 19104, USA
e-mail: MBrianBlake@drexel.edu

© Springer International Publishing AG 2018
R. Lee (ed.), *Software Engineering Research, Management and Applications*,
Studies in Computational Intelligence 722, DOI 10.1007/978-3-319-61388-8_6

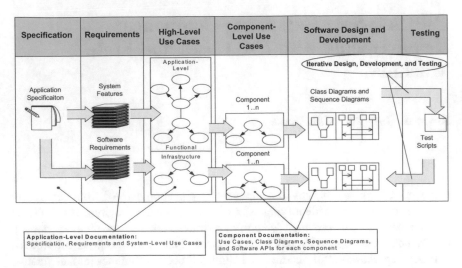

Fig. 1 The C-PLAD model [2]

the preliminary system description is written in the form of a problem statement, jointly crafted by the domain expert and the software engineers. In the Requirements Phase, software engineers again consult with the domain experts to draft a set of written requirements that separates system features from functional requirements. During a High-Level Use Cases Phase, specific C-PLAD use case templates are employed to characterize the application with a separation of functional and non-functional application concerns. Although we direct readers to related work for details [2], this approach is the first major departure from other approaches such as (the RUP and Object Modeling Technique (OMT) [15]). The Component-Level Use Cases Phase provides a step-wise iterative process for extracting component-level use cases from the high-level use cases in the previous phase (an adaptation of this process is discussed in later sections and shown in Fig. 2). The last two phases, System Design and Development and Testing follow conventional notions of use-case driven development [13] performed iteratively to help enhance use cases as more concrete notions of the target application are discovered.

The C-PLAD [2] approach provided a unique design framework facilitating the conceptualization of components. C-PLAD's major innovation was a specific step-wise process for creating use cases that support both application-level and domain-level analysis. This approach introduces new specific use case templates to assist this step-wise process. This step-wise process ultimately results in component-based use cases. In our original presentation of C-PLAD, we suggested following standard RUP techniques for design and implementation. In this paper we revise this recommendation to suggest state machine models at the component level.

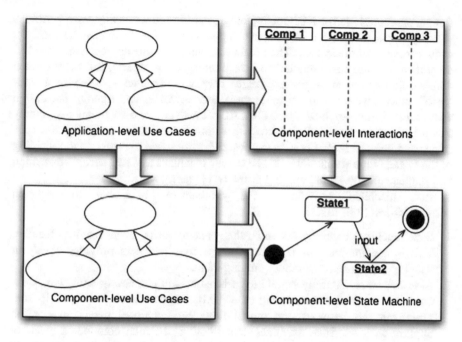

Fig. 2 C-PLAD extended into the design space

2 Component Use Cases

In C-PLAD a component-level use case is extracted from high-level use cases with other components serving as actors triggering the use case (Fig. 1). Component-level use cases are then elaborated with conventional modeling notations such as class and sequence diagrams. Initially we focused on design for relationships and interactions between components. What we have found in practice is that state machines, applied at the component level, provide an excellent way to bridge the space from use case component requirements to component design and implementation. In the following sections we describe the mapping between these spaces, present a concrete example of C-PLAD applied in an open source toolkit in a mission critical domain, and discuss how this C-PLAD extension benefits component-based software architectures.

2.1 Components and State Machines

Our initial approach to realizing application functionality using C-PLAD considered interactions between components as links representing composed component functionality. This is a fairly typical approach when performing component-based design, particularly in the RUP methodology: partition functionality, map partitions to components, then determine how to compose components to realize the entire set of functionality of the system. The emphasized tools[1] in this approach are the class diagram to capture static relationships between types of objects, and the sequence diagram, to capture messages exchanged between components to realize a "scenario". Indeed we have found these tools very useful in constructing the design abstractions in our application of C-PLAD. However, we also found that simply treating components as "black-boxes", and relying on human judgment to determine what belongs in that box, was not satisfactory when the safety and reliability of the system were at stake. We turned to state machines as a key design abstraction and implementation model applied at the component level.

State machines provide several mechanisms amenable to realizing component-level use cases:

1. State machines naturally decouple the generator of an event from its handler. This is a common feature of reactive systems, and here provides a way of "wiring" components without coupling them.
2. State machines naturally handle asynchronous and concurrent situations. Concurrency may be modeled using other UML diagrams such as an activity diagram, but this assumes one model from the functional perspective at the application level. State machines define how a particular component reacts in response to potentially numerous threads of control at a given time.
3. State machines provide a model of how components handle complexity without necessarily violating component encapsulation. Certainly to some extent, a state machine specified at the component level reveals information about the component's dynamic behavior; yet it still does not have to reveal how that behavior is implemented.
4. Components are typically expressed at multiple levels of granularity within a given system. Components are often aggregated (or composed) into higher-order components. Hierarchical state machines (HSMs) support nesting structures that may be mapped to component aggregation hierarchies.
5. State machines are unique in that they express a model of the system from the component or object perspective. As such, state machine models, when expressed per component, express information about component state and dynamic behavior that is not wholly present when using an interaction diagram.
6. State machines provide a consistent pattern that guide not only design but also visualization and implementation of a component. In the IGSTK project described below, the state machine drives component design and also component implementation. All stakeholders, no matter what area of specific expertise, share a common language for expressing how their components behave.

---

[1]We acknowledge the reader may prefer other tools, such as component instead of class diagrams, and collaboration instead of sequence diagrams. The focus is on static and dynamic relationships between components.

Our focus on component design led us to state machines. This is not altogether surprising. The principal contribution of this C-PLAD extension is the connection between component-level use cases and component-level state machines.

2.2 Extending C-PLAD with State Machines

We extend C-PLAD to emphasize the creation of state machine models at the component-level, and the connection of infrastructure functionality expressed in component-level use cases to these state machines. Further, we map component interactions to entry points in state machines for components. We show the extension to the original C-PLAD approach in Fig. 2. This extension is applied at the middle tiers section of Fig. 1 labeled "High-level Use Cases", "Component-level Use Cases", and "Software Design and Development".

In C-PLAD-SM, each component-level use case is realized by a component, and each component has a state machine capturing the encapsulated functionality of that component. This has the effect of completing the design perspective of the system; interactions and relationships are represented using sequence and class diagrams, component behaviors using state machines. Class and sequence diagrams capture static and dynamic relationships between components; state machines capture behavior from the component perspective. One can now go directly to the state machine of a component to understand if new or modified requirements will impact the system's components in an adverse way. In fact, we believe it is appropriate to construct the state machine model of the component first and use this to drive the development of the interface (class) and available collaborations (sequence) in which the component may fruitfully participate.

The trigger action of an application scenario causes an entry into a component. This "entry" is an event or request on the component. The component leverages the state machine to determine if it can safely respond to the request. Typically, a component will interact with other components to satisfy the request. In C-PLAD-SM these interactions are expressed using standard UML sequence diagrams. An interaction from some component A to some component B captured in a sequence diagram results in a request on B. Component B's ability to respond to the request is governed by its state machine (see Fig. 3). Note that in order to satisfy A's request, B might in fact have to enact complex behaviors or even delegate to other components. Instead of indicating these behaviors via self-directed messages on the sequence diagram, we use the component state machine to express this behavior.

Readers experienced in UML and component-based software will not find this result surprising; interaction diagrams focus on the collaborative behavior of a set of components to satisfy some request, state machines focus on the states and behaviors of components (or objects). Referring again to Fig. 2, what is interesting in C-PLAD-SM is that the component-level use cases are used to derive the component state machine. As the component use cases are derived from application

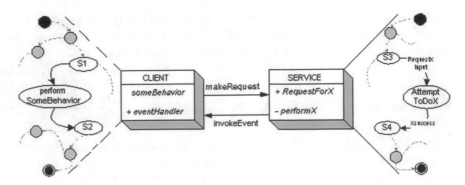

Fig. 3 Component interactions

requirements, this provides a direct path from application-level requirements to stateful behaviors on components. Furthermore, the path from application scenarios to sequence diagrams and onto component state machines also defines another path that connects application functionality to component behavior.

3 Applying the Method: IGSTK

The image-guided surgery toolkit (IGSTK, http://www.igstk.org) is an open source project aimed at developing robust software for medical applications [7]. Image-guided surgery involves the use of pre-operative medical images to provide image overlay and instrument guidance during procedures. Image-guided surgery systems have been commercially available for about 10 years now, but this field of research is still active, and challenges still exist. These systems are software intensive and must be reliable since they are used in a surgical environment.

An example application for IGSTK is an ultrasound-guided biopsy. Requirements for this application were elicited from clinicians at a major University Medical Center. The activity diagram in Fig. 4 shows the main success scenario of this application:

IGSTK employs a component-based architecture in a layered architecture pattern (see Fig. 5). An IGSTK application, shown at the left, interacts with View components that determine how to render Spatial Object Representations. These Representations are mappings of Spatial Objects, which are wrappers for objects in the surgical environment, such as needles, imaging devices, and so on. The position of objects in the space is correlated via tracking devices. A full description of the IGSTK architecture can be found in [7].

Each layer in Fig. 5 is dedicated to a type of general-purpose functionality, such as Viewing or Tracking. Each layer is composed of one or more components that realize functionality mapped to that layer. For example, consider a magnetic or optical tracking device. These devices are responsible for identifying the precise

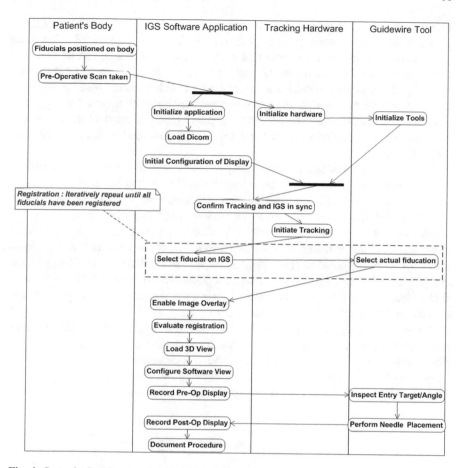

Fig. 4 Scenario for ultrasound-guided biopsy procedure

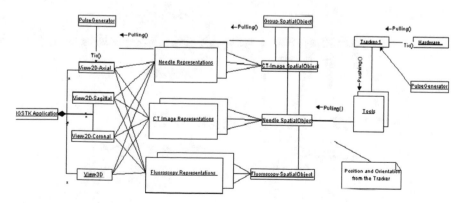

Fig. 5 IGSTK architecture layers and components

location of a surgical instrument or anatomical object (e.g. organ). The software interface of such a device is presented to IGSTK as an instance of a Tracking component. It is important to note that each component instance has its behavior regulated by a distinct, hidden state machine instance. The state machine is explicit but fully encapsulated within the component implementation; other component instances are not aware of its existence nor can they dispatch events directly to the state machine. All events are received by strongly typed component interfaces and subsequent invocation of an appropriate behavior is governed by the state machine (see Fig. 3).

Figure 5 shows associative relationships for Trackers with other components as well as interaction paths with these components. It says nothing about the behavioral aspects of the Tracker itself. Likewise, application requirements such as those shown in Fig. 4 assume robustness of various functionalities provided by the component, such as "Initiate Tracking", which is a non-trivial process. State machines were incorporated as a base architectural pattern early in the implementation cycles of the project. These state machines are applied principally to provide component safety through self-awareness; applications built on IGSTK are assured that components are explicitly aware of their own state and the events they are capable of safely responding to at any given time.

To further clarify how the C-PLAD-SM extension applies to IGSTK, we consider the Tracker component in more detail. The scenario expressed by the activity diagram in Fig. 4 describes application-level functionality. This functionality is incorporated into a high-level use case indicating the functional need to include tracking. Other high-level use cases express a similar need. Component-level use

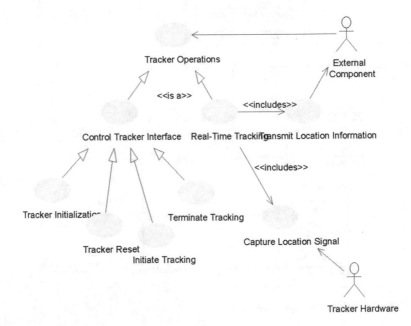

Fig. 6 Tracker component use cases

cases capturing the internal behaviors of Tracking components are constructed. These use cases and their relationships are shown in Fig. 6.

The functionality described by Fig. 6 deals with the complex behaviors (note Tracker Initialization is included in the component use cases) particular to wrapping a Tracking device as a software component. This functionality does not have to do with the interactions of Trackers with other components, except where indicated by the presence of the External Component actor.

The state machine for the Tracker component must then capture these component-specific behaviors, and when these behaviors may be invoked. The state machine regulating the behavior of a Tracker component in IGSTK is shown in Fig. 7. Note how the states and transitions at the top of the diagram correspond to the initialization activity. This model still does not indicate how this behavior is performed, but it does indicate when it may be performed in terms of the state the Tracker must be in and the event(s) it must receive.

The Tracker component state machine is interesting to review as it shows complex behavior of the IGSTK framework encapsulated within individual components housed in separate layers of the architecture shown in Fig. 5. The requirements at this component-level were derived from high-level use cases at the application level and the functionality mapped onto components via state machines.

The IGSTK project incorporates lightweight, agile-like process practices [4] into an open source project. The requirements, architecture, and component implementations have evolved iteratively over the past 2 years. The original C-PLAD method was applied in the early stages of the project to help identify components. As the project has evolved, C-PLAD-SM has also evolved.

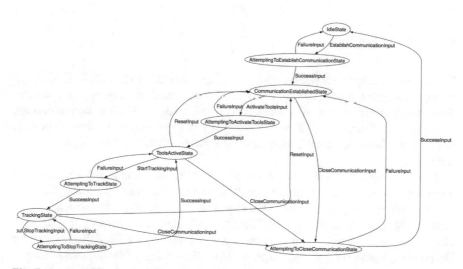

Fig. 7 State machine for IGSTK tracker component

4 Benefits of C-PLAD-SM

Extending C-PLAD component-level use cases by applying state machines as the corresponding modeling tool in the design space provides several benefits, which we discuss in this section.

4.1 Safety

Clearly, in a mission-critical domain such as software for technology-assisted surgical intervention, application safety is a crucial issue. We conceptually define safety with respect to IGSTK as predictable behavior of software components. This definition does not assume a lack of hardware or software faults, but instead states that even in the face of such faults the component remains "self-aware", understanding its current state and what behaviors it may still safely perform.

Component-level safety in IGSTK is provided by the state machine. The consequences of entering a state which represents a malfunction on the part of the component are documented in component-level use cases and reified in the component-level state machine. This is particularly important in surgical intervention scenarios, as it provides a direct route by which a critical fault in a given component may result in a drastic recovery action, such as entering a fail-safe mode where the computer is immediately shut off, the application as a whole disengaged, and human intercession is required to complete the procedure.

4.2 Visibility

To maintain component safety, state machine instances within IGSTK are constructed programmatically, as opposed to common techniques using code generation and round-trip engineering tools. This forces state machines to be constructed in a type-safe manner. IGSTK makes heavy use of C++ macros to create type-safe interactions between components and the state machines that regulate their behavior.

Hand-coding state machines with IGSTK has the unfortunate side-effect of not allowing component designers to "see" what the state machine looks like. To provide this facility, IGSTK includes export methods to multiple formats so that state machines may be visualized. These visualizations may then be inspected to ensure proper behavior. Visualization formats include .dot files displayable by GraphViz (http://www.graphviz.org), and .lts files displayable by LTSA (http://www.doc.ic.ac.uk/~jnm/book/). We are currently constructing more general export routines to support UML's XML interchange format (XMI) and the W3C's statechart SCXML format (http://www.w3.org/TR/scxml/).

Using animation techniques with LTSA, we have added the ability to trace the execution of an IGSTK application by showing the sequence of transitions through which each state machine progresses on a per component basis. This type of record-playback facility, derived from IGSTK log files, allows component developers to ensure per component behaviors are sequenced correctly, and application architects to ensure cross-component interactions occur as expected.

4.3 Maintainability

The connection between a component-level use case and a component-level state machine provides perhaps its greatest benefits in the area of software maintenance. In fact, this is one of the strongest situations supporting the use of C-PLAD in general. Experienced architects are able to successfully digest requirements and from scratch produce a component-based software architecture that satisfies those requirements. It is more challenging, in our view, to maintain the consistency and intent of the architecture in the face of new and evolving requirements. When the scenarios supporting a use case are modified, how does one navigate through the original analysis and design to understand the impact to the developed architecture. The obvious answer is "very carefully", as the architecture may easily become misunderstood and bloated over time.

C-PLAD provides a process whereby requirements traceability from the application-level to the component-level is provided via the respective use cases, and traceability into the component design is preserved by mapping from the component-level use case to the component state machine.

Consider the following situation. Suppose a new scenario is proposed that suggests a modification to a component-level use case. The provisions of these changes must be incorporated into the component's state machine. If significant redesign of the state machine is required (new states, new transitions, refactoring of existing transitions), it suggests the new scenario was not properly mapped to the component architecture. Perhaps a new component is required, preventing architecture bloat at the component-level, or the component design is too brittle and should be revisited. Stated another way, component state machines are an expression of the functional purpose of the component, providing a mechanism for maintenance analysis that goes beyond human subjective judgment of the relationship of new or modified functionality to the existing components. The state machine provides a tangible means by which one can assess the impact of evolving requirements on the evolution of the component architecture.

4.4 Verifiability

Traceability is well understood to provide maintainability and testability of a software platform. The mapping of component-level use cases to component state machines not only assists with maintainability but with the ability to validate the components provide the functionality required according to the component-level use cases. Originally we considered applying well-established model-checking tools such as SPIN (http://spinroot.com) and UPPAAL (http://www.uppaal.com) to this task. We found these tools are geared toward verifying properties of state machines for reactive concurrent systems, and IGSTK component state machines require a more safety-oriented approach as described above. We have constructed [8, 9] a suite of validation tools that verify certain safety properties of component state machines and considering ways to automatically validate a component provides all the behaviors mandated by its component-level use cases.

5 Related Work

The initial goal of C-PLAD was to couple the creation of software product lines [11] with the integration of openly available components [1] to address what Metzger and Pohl now call "Variability Management" [14] in software product-line engineering. However, another feature in this work is the extension of C-PLAD to support the formal integration of state machine paradigms within component software. Several places explore low-level semantics of state machine design (c.f. [6, 10]), but few research projects investigate the incorporation state machines into formal software development lifecycles. Furthermore, we focus on the integration of state machine software, which is different than the common approach of modeling and evaluating software using state chart diagrams offline (at design time or during testing) [8]. Here, we intend to conceptualize the functional requirements needed then develop/generate state machine software to ensure that functionality. Bontemps et al. [3] investigates the generation of agent-oriented software from state-based models but mainly for synthesis and verification. In our work, we develop state machines that control software in actual operational environments. Zhu [16] directly ties scenario descriptions to behavioral component specifications for an agent-oriented platform. This approach is similar to the tack described in this chapter, though our focus is on the internal safety per component and its correlation to a component-level use case, as opposed to component interactions to satisfy an application-level scenario. Coleman et al. [5] introduce ObjectCharts that incorporate state machines into the object-oriented development lifecycle. The focus of this work is similar, particularly with regards to the consideration of software lifecycles. However, ObjectCharts are focused at the class-level, where as our work addresses component-level design which resides at an aggregate level. In a recent survey by Khan et al. [12] on non-functional requirements engineering, the authors provide a taxonomy for evaluating integrating software architectures with non-functional requirements approaches together with notations used in each approach. C-PLAD-SM integrates multiple notations and techniques to address functional and non-functional requirements in component-based architectures.

6 Conclusions

Component encapsulation provides several features for managing and evolving complex software. Components typically provide well-defined interfaces representing a rigid boundary between providers and consumers of services. Components provide a pragmatic vehicle for testing and maintaining complex software. Component-based design, and specifically encapsulation, does not apply as well to other design needs such as dealing with per component complexity. While partitioning the complexity of a software system into components helps make the overall complexity of the system manageable, it is still the case that many of these "smaller problems" remain complex and require one to consider how to express object-level realization of this complex functionality. Additionally, many types of application domains, including the one in which we have applied C-PLAD-SM, require stronger statements of the dynamic behavior of a component in order to guarantee quality attributes such as safety and maintainability. Our solution is to construct a state machine, derived from the component-level use case, at the component level.

In the conclusion of our original presentation [2], we suggested C-PLAD as a technique that bridges the gap between RUP's use case driven requirements methodology and (human-driven) derivation of an architecture. Few methodologies address this space, instead, it is up to talented individuals with the experience and insight to produce powerful abstractions to drive solutions. We do not suggest C-PLAD replaces this innate talent, but we do put it forward as an extension to RUP that attempts to address the area directly. In fact, our approach, is consistent with the evolution of UML—providing multiple perspectives on a problem to arrive at a solution. In the case of C-PLAD-SM, we use the application and the domain level for requirements, the high-level and the component use case for analysis, and class, sequence and state machine perspectives of components at the design level to arrive at an architecture.

References

1. Batory, D., Johnson, C., MacDonald, B., von Heeder, D.: Achieving extensibility through product-lines and domain-specific languages: a case study. In: ACM TOSEM, April 2002
2. Blake, M.B., Cleary, K., Ibanez, L., Ranjan, S., Gary, K.: Use case-driven component specification: a medical applications perspective to product line development. In: Proceedings of the ACM Symposium on Applied Computing (SAC'05), Software Engineering Track, Sante Fe, New Mexico, Mar 2005
3. Bontemps, Y., Heymans, P., Schobbens, P.-Y.: From live sequence charts to state machines and back: a guided tour. IEEE Trans. Softw. Eng. **31**(12), 999–1014 (2005)
4. Campanelli, A., Parreiras, F.: Agile methods tailoring–a systematic literature review. J. Syst. Softw. **110**, 85–100 (2015)
5. Coleman, D., Hayes, F., Bear, S.: Introducing objectcharts or how to use statecharts in object-oriented design. IEEE Trans. Softw. Eng. **18**(1), 1992

6. Ferrentino, A.B., Mills, H.D.: State machines and their semantics in software engineering. In: Proceedings of the IEEE Conference on Computer Software and Applications (COMPSAC'77), pp. 242–251 (1977)
7. Gary, K., Blake, B., Ibanez, L., Gobbi, D., Aylward, S., Cleary, K.: IGSTK: an open source software platform for image-guided surgery. In: IEEE Computer Special Issue on Software Engineering Systems, April 2006
8. Gary, K., Kokoori, S., David, B., Otoom, M., Cleary, K.: Architecture validation in open source software. In: Proceedings of ROSATEA 2007: The Role of Software Architecture for Testing and Analysis, Boston, MA, July 2007
9. Gary, K., Kokoori, S., Muffih, B., Enquobahrie, A., Cheng, P., Yaniv, Z., Cleary, K.: Agile methods for safety-critical open source software. J. Softw. Pract. Exp. (2011) (Wiley)
10. Harel, D., Naamad, A.: The STATEMATE semantics of statecharts. ACM Trans. Softw. Eng. Method. (TOSEM) 5(4), 293–333 (1996)
11. Heineman, G.T., Councill, W.T. (eds.): Component-Based Software Engineering: Putting the Pieces Together. Addison-Wesley, Boston, MA (2001)
12. Khan, F., Jan, S.R., Tahir, M., Khan, S., Ullah, F.: Survey: dealing non-functional requirements at architecture level. VFAST Trans. Softw. Eng. 9(2), 7–13 (2016)
13. Kruchten, P.: The Rational Unified Process—An Introduction, 2nd edn. Addison-Wesley, Boston, MA (2000)
14. Metzger, A., Pohl, K.: Software product line engineering and variability management: achievements and challenges. In: Proceedings of the on Future of Software Engineering, pp. 70–84. ACM (2014)
15. Rumbaugh, J., Blaha, M., Premerlani, W., Eddy, F., Lorensen, W.: Object-Oriented Modeling and Design. Prentice Hall (1991)
16. Zhu, X., Maiden, N., Pavan, P.: Scenarios: bringing requirements and architecture together. In: Proceedings of the 2nd International Workshop on Scenarios and State Machines: Models, Algorithms, and Tools (SCESM'03). Portland, OR (2003)

A Structural Rule-Based Approach for Design Patterns Recovery

Mohammed Ghazi Al-Obeidallah, Miltos Petridis and Stelios Kapetanakis

Abstract Design patterns have a key role in the software development process. They describe both structure, behavior of classes and their relationships. Design patterns can improve software documentation, speed up the development process and enable large-scale reuse of software architectures. This paper presents a Multiple Levels Detection Approach (MLDA) to recover design pattern instances from Java source code. MLDA is able to extract design pattern instances based on a generated class level representation of an investigated system. Specifically, MLDA presents what is the so-called Structural Search Model (SSM) which incrementally builds the structure of each design pattern based on the generated source code model. Moreover, MLDA uses a rule-based approach to match the method signatures of the candidate design instances to that of the subject system. As the experiment results illustrate, MLDA is able to extract 23 design patterns with reasonable detection accuracy.

Keywords Design patterns · Detection · Reverse engineering · Gang of four · Static analysis · Rule-based systems

1 Introduction

The detection of design patterns is a reverse engineering activity where design patterns are extracted depending on certain criteria. The idea of patterns was adopted by the so-called Gang of Four (Gamma, Helm, Johnson, and Vlissides) [1]-henceforth GoF.

M.G. Al-Obeidallah (✉) · S. Kapetanakis
Department of Computing, University of Brighton, Brighton, UK
e-mail: M.Al-Obeidallah@brighton.ac.uk

S. Kapetanakis
e-mail: S.Kapetanakis@brighton.ac.uk

M. Petridis
Department of Computing, Middlesex University, London, UK
e-mail: M.Petridis@mdx.ac.uk

© Springer International Publishing AG 2018
R. Lee (ed.), *Software Engineering Research, Management and Applications*,
Studies in Computational Intelligence 722, DOI 10.1007/978-3-319-61388-8_7

107

GoF have cataloged 23 design patterns. Each design pattern describes a problem that occurs over and over again, in an attempt to describe the core solution to that problem. This solution can be re-used a million times over, without doing it the same way twice. In fact, design patterns vary in their levels of abstraction. Each design pattern solves a specific design problem by connecting a number of classes (participant classes) together using different relationships. According to the GoF's catalog, each design pattern involves both structural and behavioral aspects. Structural aspects describe the static arrangement of classes and their relationships. On the other hand, behavioral aspects describe dynamic interactions between the participant classes.

Design patterns at the source code level reflect the earliest set of design decisions that have been taken by the development team. In addition, the majority of current software systems involve instances of design patterns in their source codes. Consequently, the extraction of design patterns helps a number of stakeholders, such as system analysts, software engineers and architects to capture design and code information and enhance their understanding over an enterprise system. However, the extraction of design patterns is not an easy task since software documentation is not always available and the possible variants of pattern instances.

This paper presents a new detection approach named Multiple Levels Detection Approach (MLDA). MLDA provides a Structural Search Model (SSM) that is able to extract the instances of design patterns based on the generated class level representation of the Java source code.

Furthermore, MLDA builds the structure of each design pattern incrementally based on the generated source code model. To reduce the number of false positive instances, a rule-based system that is able to match the method signatures of the candidate design instances to that of the subject system has been developed. Since GoF design patterns have been attractive both to industry and academia, this paper focuses on the GoF design patterns.

This paper is organized as follows: Section 2 presents MLDA architecture, the structural search model and the rule-based system. Experiments and results, related work and threats to validity are discussed in Sects. 3, 4, and 5 respectively. Finally, the conclusion is presented in Sect. 6.

2 Recovering Design Pattern Instances

The Multiple Levels Detection Approach (MLDA) is a research prototype, which has been developed to extract the instances of design patterns from Java source code. MLDA involves three main levels: a parsing level, a searching level and a method signatures matching level. The architecture of MLDA appears in Fig. 1.

The parsing level aims to extract the source code information and produce a source code model. Moreover, MLDA aims to extract the five major relationships, which may occur between classes and objects inside any object-oriented program. These relationships are Inheritance, Aggregation, Association, Dependency and Realiza-

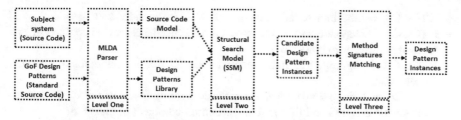

Fig. 1 The architecture of MLDA

tion. On the other hand, the searching level of MLDA aims to examine the source code model that has been developed during the parsing level and tries to match it with the GoF's catalog.

Specifically, MLDA introduces a structural search model (SSM), which involves a searching algorithm for each design pattern. MLDA works on the principle of building the patterns incrementally based on the connecting relationships. The third level of MLDA is the method signatures matching level. The method signatures of the subject system are represented as a set of facts. On the other hand, the required method signatures of the candidate design instances are represented as a set of rules. CLIPS (C Language Integrated Production System) [2], an expert system tool, has been used to match the generated facts and rules. It must be noticed that MLDA uses the standard codes presented by GoF [1].

2.1 Parsing Level

Parsing is "the process of analyzing a string of symbols, either in natural language or in computer languages, conforming to the rules of a formal grammar" [3]. MLDA's parsing level relies on the packages of the Javaparser version 1.0.11, which has been developed by Júlio Vilmar Gesser and is available online [4]. The Javaparser is an open source project and can be used under the terms of the LGPL license. In fact, the Javaparser involves a number of useful packages, such as Japa.parser, Japa.parser.ast, Japa.parser.ast.expr and Japa.parser.ast.visitor. The motivation of importing the packages of Javaparser is their ability to generate an Abstract Syntax Tree (AST) that can record the source code structure. AST is a tree that represents the syntactic behavior of the source code where its elements are mapped into tree nodes. The parsing level of MLDA aims to extract all the possible relationships between classes in the Java source code. The relationships have been defined by MLDA as follows: The relationship between the source class Cs and the destination class Cd is denoted by R (Cs, Cd) = Inheritance, Dependency, Aggregation, Association, and Realization, in which:

- R (Cs, Cd) = Inheritance: indicates that class Cd is extended by class Cs.
- R (Cs, Cd) = Dependency: indicates that a reference to class Cd is passed in as a method parameter to class Cs.
- R (Cs, Cd) = Aggregation: indicates that class Cs stores a reference to class Cd for later use.
- R (Cs, Cd) = Association: indicates that the containing object in class Cs is responsible for the creation and lifecycle of the contained object of class Cd.
- R (Cs, Cd) = Realization: indicates that class Cd is an interface and extended by class Cs.

MLDA will try to extract all classes and their relationships from the subject system. In fact, MLDA provides a clear distinction between the aggregation relationship and the association relationship. In the aggregation relationship, the creation of the objects will occur during the compile time while in the association relationship, the creation of the objects will occur during the runtime. Hence, MLDA performs a static analysis that is acting as a dynamic analysis since all the object creations are recorded. In addition, the parsing level of MLDA records the dynamic interactions between classes.

The output of the parsing level is a model of the source code, and a library of design patterns. The source code of the subject system is modeled in the form: source class, destination class and relationship type. The same structure is also applied to represent the catalog of GoF. The source code model generated by MLDA's parsing level is presented in Fig. 2.

The source code model will be exported into an SQL table, which will be examined by the SSM in order to extract the candidate instances of design patterns. The library holds a representation of each design pattern. This representation is similar in its structure to the structure of the source code model (i.e. the representation of

Fig. 2 The source code model generated by MLDA

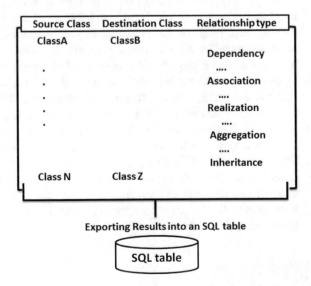

Class_Relations		
Source_Class	Destination_Class	Relation_Type
ConcreteCommand	Receiver	AGGREGATION
ConcreteCommand	Command	INHERITANCE
Invoker	Command	AGGREGATION

Fig. 3 The representation of the command design pattern in the library

each design pattern in the library involves three columns: source class, destination class and relationship type). To explain how MLDA represents each design pattern in the library, the command design pattern representation is presented in Fig. 3.

MLDA has successfully extracted two aggregation relationships and one inheritance relationship. One aggregation relationship is connecting the "Invoker" class to the "Command" class, and the other is connecting the "ConcreteCommand" class to the "Receiver" class. Furthermore, the inheritance relationship between the "Command" class and the "ConcreteCommand" class is also extracted. However, MLDA has excluded the role of the "Client" class since it represents the role of the main program inside the source code. This will not affect how the command participant classes are connected and communicate together.

2.2 Searching Level

The searching level of MLDA aims to build the design pattern structure incrementally from the source code model based on its representation in the library. MLDA involves a searching algorithm for each design pattern.

The searching algorithm tries to build the pattern structure from the source code model by checking the relationship that is connecting the source class to the destination class. If the search process finds one of the required relationships of the pattern, it will continue searching for the remaining relationships until it can form a complete pattern structure that is similar to the pattern representation in the library. When the pattern structure has been found in the source code model, all the patterns' participant classes are exported from the source code model to an SQL table. MySQL Workbench version 6.3 CE was used to create the tables. Since MLDA is able to distinguish between the aggregation and the association relationships and records all the object creations and the dynamic interactions between classes, it is expected to extract all behavioral design patterns. In addition, the searching level of MLDA introduces what is called the SSM which involves a searching algorithm for each GoF design pattern. Each part of the SSM involves two participant classes (i.e. the source and destination class).

To explain how the searching level works, the searching algorithms of the Proxy and Command design patterns are presented. Figure 4 shows the SSM of the Proxy design pattern. The extraction of Proxy instances is based on its representation in

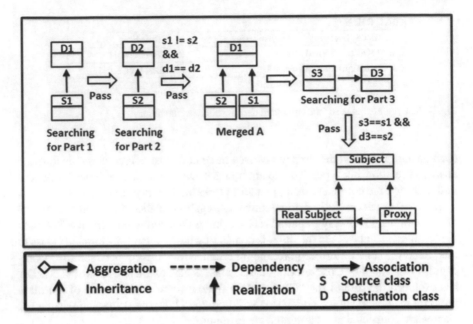

Fig. 4 The structural search model of the proxy design pattern

the library. MLDA will search for two participant classes that have a realization relationship. The retrieved classes are stored temporally for later use. Then, MLDA will
continue searching in the table, which represents the source code model, for another
two classes that are connected together by a realization relationship as well. MLDA
will combine the two extracted parts together if they have the same superclass (root)
and different subclasses.

If MLDA has successfully combined the two extracted parts, the process will continue searching for another two classes that are connected together using an association relationship. All classes are forming an instance of the Proxy design pattern
if the role of the third part's source class is similar to the role of the merged part's
source class. In addition, the role of the destination class of the third part must be
similar to the role of the source class of the first part. All the extracted instances and
their participant classes will be exported to the SQL table.

Figure 5 explains the searching attempts to extract the instances of the Command
design pattern. The Command design pattern involves four main roles: Invoker,
Command, ConcreteCommand and Receiver. MLDA will start searching for two
classes that are connected using an aggregation relationship (searching attempt one).
The next searching attempt aims to extract another two classes that are connected
together using an inheritance relationship (searching attempt two). MLDA will combine the parts of searching attempt one and searching attempt two together if they
have the same destination class. Consequently, forming the "Merged A" part. Finally,
MLDA will search for another two participant classes that are connected using an

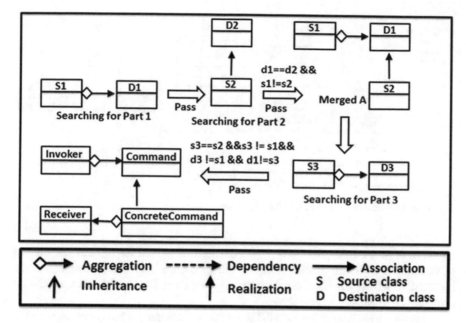

Fig. 5 The structural search model of the command design pattern

aggregation relationship and which differ from the classes that are extracted during the third searching attempt. If the second and third searching attempts have the same source class, then all classes together form an instance of the Command design pattern.

The design pattern library generated by MLDA and the structural search model and its pseudocode for all GoF design patterns are available online and can be downloaded from https://www.sites.google.com/site/mldamodel/.

2.3 Method Signatures Matching

The candidate design pattern instances that have been detected by SSM are filtered by applying a rule-based approach. This approach aims to remove the false positive instances by matching the method signatures of the candidate design instances to that of the subject system. A rules template for GoF method signatures has been created to reflect the required method signatures for each design pattern. The rules template relies on standard structural definitions presented by GoF [1]. In addition, the so-called MLDA rules/facts generator is introduced here, which is a simple Java program that is able to write a set of rules and facts based on the method signatures representation of the candidate design instances and subject system. Specifically, MLDA rules/facts generator will generate a list of rules to reflect the required method

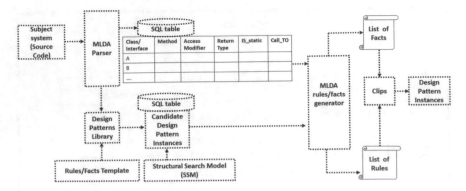

Fig. 6 A detailed architecture of MLDA's level three

signatures and method calls between candidate instance participant classes. On the other hand, MLDA rules/facts generator will generate a list of facts to represent the interactions between methods inside the subject system. Figure 6 shows the detailed architecture of MLDA's level three.

A rule based-system contains IF-THEN rules, facts and an inference engine that controls the application of the rules. Our main motivation behind the use of a rule-based approach is the ability to represent the method signatures of the candidate design instances as an independent piece of knowledge, which can be transformed into a set of rules. In addition, the method signatures of GoF design patterns have a uniform structure, which facilitate their representation as a set of rules. By contrast, the comparison process performed by the inference engine allows an effective match between the set of rules and the facts. MLDA uses CLIPS v6.3, an expert system tool, to process the generated facts and rules and to remove the false positive instances. The detailed architecture of MLDA illustrates the relationship between the SSM, MLDA rules/facts generator, rules template and CLIPS.

As Fig. 6 illustrates, MLDA parser will parse the subject system and extract its method signatures from each class/interface. The extracted signatures are access modifier, is_static, returntype and call_to.

2.3.1 CLIPS

CLIPS is one of the most popular shells widely used through the government, industry and academia. The CLIPS shell provides the basic elements of a rule-based system:

- Fact-list, which contains all the facts about the problem. Facts are stored in short-term memory.
- Knowledge-base, which contains all the rules. Rules are stored in the knowledge base (database).
- Inference engine, which controls the overall execution of rules.

CLIPS uses forward chaining and provides a language for representing facts and rules. The language is based on the artificial intelligence language, LISP. CLIPS inference engine enacts the required matching by using Rete algorithm [5]. Rete algorithm is a pattern matching algorithm for implementing expert systems designed by Charles Forgy in 1974 [5]. It is used to determine which rule the inference engine should fire.

2.3.2 Rules Template for Method Signatures of Design Patterns

A rules template has been created to reflect the required method signatures between pattern participant classes. A readable and uniform rule syntax were used, which is consistent with the CLIPS rules' syntax. Hence, the generated rules can be loaded directly onto CLIPS. Table 1 shows the rule syntax used to create the template and its corresponding significance. Table 2 shows the rules template of the proxy design

Table 1 The created rules template syntax and its significance

Rule syntax	Significance
Class A has method m	Method m is implemented inside Class A
Method m returntype Class A	Method m returns an object of type Class A
Method m is static YES/NO	Whether method m is static or not
(test (=(str-compare m1 m2)0)	To check whether m1 and m2 are the same method (for overriding purpose)
(test (neq m1 m2))	To check whether m1 and m2 are two different methods
(method m1 call_ to method m2)	The implementation of method m1 involves a call to method m2

Table 2 Rules template of the proxy design pattern

	Proxy rule template
1.	(defrule Proxy_rule
2.	IF
3.	(Subject has method ?x)
4.	(Proxy has method ?y)
5.	(test (= (str-compare ?x ?y) 0))
6.	(RealSubject has method ?z)
7.	(test (= (str-compare ?x ?z) 0))
8.	(method ?y call_ to method ?z)
9.	THEN
10.	(Prox_ instance is true positive)
11.	) End of proxy rule

pattern. The rules template for all GoF design patterns is available online and can be downloaded from MLDA website.

2.3.3 MLDA Rules/Facts Generator

MLDA rules/facts generator-henceforth R/F generator-is a simple Java program, implemented as a part of MLDA project, which generates a set of rules and facts to represent the required method signatures of candidate design instances and a subject system respectively. The outputs of the R/F generator consist in two files: rules.txt and facts.txt, which will be loaded onto CLIPS for processing.

In order to generate the rules, the R/F generator constructs a connection to the SQL table and provides access to each record. Based on the rules template, the R/F generator will generate a rule for each candidate design instance. Specifically, the R/F generator will fill each template entry by its corresponding role in the SQL table. Java OutputStreamWriter has been used to write rules into a text file.

Figure 7 presents an example to illustrate how the R/F generator creates rules. The presented example shows the generated rules of the Proxy candidate instances.

As Fig. 7 illustrates, the R/F generator creates two rules to represent the required method signatures between Proxy participant classes (that is, Subject, Proxy and RealSubject). Method IDs, titles and variables will be automatically incremented as new instances are inserted into the table. To generate the facts, the MLDA parser stores the retrieved method signatures from the subject system in an SQL table. The R/F generator represents each method signatures record of the subject system as a set of facts. The subject system is represented as a set of classes/interfaces where each record stores the methods that are implemented inside that class/interface. In addition, each record stores the method access modifier, return type, static status and method calls. Figure 8 shows an example of how the R/F generator generates

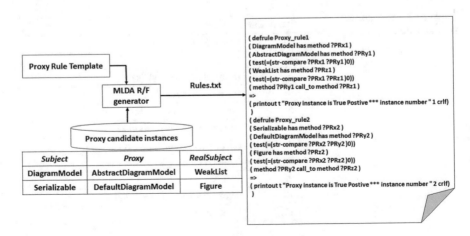

Fig. 7 Rules generation example of the proxy candidate instances

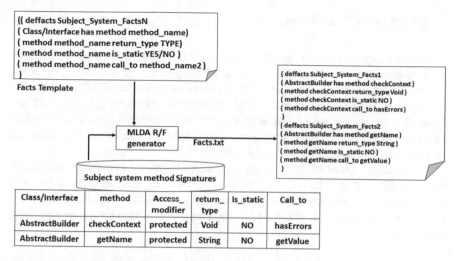

Fig. 8 Facts generation example of subject system

facts to represent the method signatures of a subject system. The R/F generator is customizable. This means that the syntax of the rules and facts can be changed by modifying the template.

2.3.4 Rules and Facts Matching

The generated facts and rules will be loaded onto CLIPS, as two text files, for processing. Facts will be stored in the working memory while the rules will be stored in the knowledge base. The CLPS inference engine uses forward chaining which relies on Rete algorithm to fire the rules. If rule conditions match a set of facts, the rule will be inserted into the agenda for execution. The agenda is a collection of activations which are rules match pattern entities. At the end of the cycle, all the matched rules will be in the agenda. The inference engine will fire the rules based on their order in the knowledge base. The topmost rule will be executed first. However, the order of the rules in the knowledge base is not important. This is mainly due to the way that the rules template was constructed. Specifically, the action part of each rule does not assert new facts to the working memory. Hence, the order of the rules will not affect rules execution. One cycle is required to fire the rules. The "run" command would run the inference engine of CLIPS. Rules that represent false positive instances should not be fired. These instances have a similar structure to the design patterns, but that they did not implement the required method signatures of design patterns. On the other hand, rules that represent the true positive instances should be fired. However, some of the true positive instances are partly implemented in the source code.

3 Experiments and Results

MLDA was implemented in Java using NetBeans Integrated Development Environment version 8.1. The extracted instances of design patterns are stored in tables, which have been constructed using MySQL Workbench version 6.3 CE. MLDA has been applied to four open source systems (JHotDraw, JRefactory, QuickUML and JUnit) that are widely used as benchmarks for design patterns detection. The characteristics of the four systems appear in Table 3. All the experiments have been run on Windows 7 with Intel Core i5-2400 CPU.

The effectiveness of MLDA has been evaluated in terms of accuracy and searching time. To evaluate the accuracy of MLDA, two well-known metrics are used namely precision and recall. The F-measure, which represents the harmonic mean of recall and precision, is calculated as well. The previous metrics can be calculated as follows [6]:

$$\text{Precision} = [\text{True Positives}/(\text{True Positives} + \text{False Positives})]\%$$
$$\text{Recall} = [\text{True Positives}/(\text{True Positives} + \text{False Negatives})]\%$$
$$\text{F-measure} = 2 \times [(\text{Precision} \times \text{Recall})/(\text{Precision} + \text{Recall})]\%$$

Where True Positives: the number of instances, which are correctly detected by MLDA. False Positives: the number of instances, which are incorrectly detected by MLDA. False Negatives: the number of instances, which are incorrectly rejected by MLDA (missed instances).

To validate the extracted instances, we refer to the all publicly published results in the literature which have presented the true positive, false positive and false negative instances of JHotDraw, JRefactory, JUnit and QuickUML. Consequently, the number of true positives, false positives and false negatives are validated based on the common public results in the literature.

The subject systems have been parsed by MLDA parser and the SSM model is applied to the generated source code model to extract the candidate instances of design patterns. The results of the parsing of the subject systems are presented in Table 4. MLDA extracted 990 classes and 80 interfaces from the subject systems. In addition, Table 4 shows the number of facts and rules generated by the R/F generator.

Table 3 The characteristics of the systems used in the experiments

Project	Category	Version	Size (MB)
JHotDraw	Graphics user interface	5.1	2.98
JRefactory	Graphics user interface	2.6.24	4.0
JUnit	Unit testing	3.7	2.66
QuickUML	Design tool	2001	1.76

Table 4 The results of the parsing and rules/facts generation of the subject systems using MLDA

System/Extracted feature	JHotDraw	JRefactory	JUnit	Quick-UML
Interfaces	18	35	8	19
Classes	183	577	104	126
Aggregation	96	439	31	99
Dependency	110	782	110	157
Association	40	54	25	118
Inheritance	97	535	50	105
Realization	25	89	13	33
Methods	572	1727	334	584
Facts	4222	9163	1981	2397
Rules	195	284	39	129

Table 5 presents the experimental results of recovering 23 GoF design patterns from the subject systems. In terms of performance, MLDA is performing quite well where it extracted 647 candidate instances within 312 s. The searching time that MLDA spent depends on the number of classes involved and the number of recovered instances. In terms of accuracy, MLDA detected most of the design pattern instances that are consistent with the standard definition presented by GoF. However, MLDA wrongly rejected some design pattern instances in the subject systems since these instances are partly implemented in the source code. For example, MLDA has rejected one visitor instance since the roles of "ObjectStructure", "Element", and "ConcreteVisitor" are not implemented in the source code of JRefactory.

MLDA builds the structure of each design pattern incrementally and records all the object creations and interactions between classes, which are necessary to detect design pattern instances. Specifically, all the object creations are recoded during the parsing level where MLDA made a clear distinguishing between the association relationship and the aggregation relationship. MLDA is not able to extract the instances that are inconsistent with GoF representation or partly implemented in the source code.

Furthermore, the rule-based approach presented by MLDA enhances the detection accuracy which relies on the principle of relationships matching. As Table 5 illustrates, the rule-based approach reduces the number false positive instances by matching the method signatures of the candidate instances to that of the subject system. The inference engine of CLIPS performs the required matching between the generated rules and facts.

The common syntax for all relationships, detected design pattern instances for all subject systems, R/F generator source code, rules template and the source code of MLDA is available online and can be downloaded from https://www.sites.google.com/site/mldamodel/.

Table 5 The experimental results of recovering 23 GoF design patterns from the subject systems

Subject systems	JHotDraw				JRefactory				JUnit				QuickUML			
Design patterns	CI	DI	P%	R%	CI	DI	P%	R%	CI	DI	P%	R%	CI	DI	P%	R%
Singleton	2	2	100	100	10	10	100	83	0	0	100	100	1	1	100	100
Prototype	2	2	100	100	0	0	NA	NA	0	0	100	100	0	0	NA	NA
Abstract factory	0	0	NA	NA	0	0	NA	NA	0	0	100	100	0	0	NA	0
Factory method	0	0	NA	0	101	81	98	91	2	2	100	100	12	12	100	67
Builder	0	0	NA	NA	0	0	NA	0%	0	0	100	100	0	0	NA	0
Adapter	31	13	85	100	17	15	100	94	7	5	100	45	26	25	100	86
Bridge	7	5	80	100	0	0	NA	NA	0	0	100	100	0	0	NA	0
Composite	1	1	100	100	0	0	NA	NA	0	0	100	0	1	1	100	100
Decorator	1	1	100	33	1	1	100	100	1	1	100	100	2	2	100	67
Facade	1	1	100	100	2	2	100	NA	0	0	100	100	1	1	100	100
Flyweight	1	1	100	100	0	0	NA	NA	0	0	100	100	1	1	100	100
Proxy	0	0	NA	NA	0	0	NA	NA	0	0	100	100	1	1	100	50
CoR	0	0	NA	NA	0	0	NA	NA	0	0	100	100	0	0	NA	NA
Command	17	9	89	89	45	32	69	88	0	0	100	100	17	17	100	94
Interpreter	0	0	NA	NA	0	0	NA	NA	0	0	100	100	0	0	NA	NA
Iterator	0	0	NA	NA	0	0	NA	NA	0	0	100	0	0	0	NA	NA
Mediator	2	0	NA	NA	0	0	NA	NA	0	0	100	100	0	0	NA	NA
Memento	1	0	NA	NA	0	0	NA	NA	0	0	100	100	6	0	NA	NA
Observer	0	0	NA	0	0	0	NA	NA	0	0	100	0	1	1	100	100
State/Strategy	33	7	86	100	11	8	100	73	8	3	100	100	11	10	100	100
Visitor	1	1	100	50	1	1	100	50	0	0	100	100	0	0	NA	NA
Template method	95	5	80	100	96	21	19	100	21	2	50	100	49	13	23	100
Total/Average	195	48	87%	84%	284	171	83%	88%	39	13	92%	57%	129	85	88%	82%

Note CI Candidate Instances after applying Structural Search Model (level two) *DI* Detected Instances after applying rules-based approach (level three) *P* Precision *R* Recall *NA* Not Applicable

4 Related Work

During the last two decades, many tools and approaches have been developed in order to recover design pattern instances from object-oriented source code. However, these tools and approaches differ in their input, output, evaluation criteria and extraction methodology.

Guéhéneuc et al. developed PTIDEJ (Pattern Traces Identification, Detection and Enhancement in Java) at the University of Montreal and since then, PTIDEJ has evolved into a complete reverse engineering tool [7, 8]. PTIDEJ extracted design patterns by finding all micro-architectures similar to design motifs (i.e. finding all classes and interfaces that have structures similar to design motifs).

Design patterns detection using a Similarity Scoring Approach (SSA) is a research prototype developed in Java at the University of Macedonia to handle the problem of multiple variants of design patterns [9]. SSA uses a graph similarity algorithm to detect design patterns by calculating the similarity of vertices between the pattern and the system under study. To handle the system-size problem, SSA divides the system into a number of subsystems and the similarity algorithm is applied to the subsystems instead of the whole system. SSA was applied to three open source systems: JHotDraw v5.1, JRefactory v2.6.24 and JUnit v3.7.

DeMIMA (Design Motif Identification: Multilayered Approach) is a tool developed at the University of Montreal [10]. It is a semi-automatic tool that identifies micro-architectures similar to design motifs in the source code. DeMIMA involves three layers: two layers to recover source code abstract model and class relationships and one layer to recognize design patterns from the abstract model. DeMIMA identifies micro-architectures similar to the design motifs by transforming them into constraints that reflect the relationships between participant classes. DeMIMA was applied to JHotDraw v5.1, JRefactory v2.6.34, JUnit v3.7, MapperXML v1.9.7 and QuickUML 2001. DeMIMA observed precision of 34% for twelve design motifs and achieved 100% recall.

The approach presented by Dongjin et al. involves a sub-pattern representation for the 23 GoF design patterns-henccforth, sub-patterns approach [11]. The source code and predefined GoF patterns are transformed into graphs with classes as nodes and the relationships as edges. The instances of sub-patterns are identified by means of subgraph discovery. The joint classes have been used to merge the sub-pattern instances. Finally, the behavioral characteristics of method invocations are compared with the predefined method signature template of GoF patterns to obtain final instances. The sub-patterns' approach achieved 68–100% precision and 73–100% recall.

Pattern Inference and Recovery Tool (PINOT) reclassifies the catalog of design patterns by intent [12]. The new classification is better suited for the reverse engineering approach. To capture program intent, PINOT used static program analysis techniques to recover design pattern instances from four open source projects: Java AWT v1.3, JHotDraw v6.0, Java Swing v1.4 and Apache Ant v1.6.

The technique presented by Uchiyama et al. (henceforth, the Uchiyama technique) uses source code metrics and machine learning to detect design patterns [13].

By using the goal question metric method (GQM), some source code metrics are selected to judge roles. Pattern specialists define a set of questions to be evaluated, and they decide on certain metrics to help answer these questions. Moreover, the Uchiyama technique uses a hierarchical neural network simulator in which the input is the metric measurements of each role and the output is the expected role. The detection is done by matching the candidate roles produced by the machine learning simulator and the pattern structure definitions. Searching is looking for all combinations of the candidate roles that are in agreement with pattern structures. The Uchiyama technique extracted inheritance, interface implementation and aggregation relationships.

The approach presented by Alnusair et al. [14]-henceforth, Sempatrec-uses ontology formalism to represent the conceptual knowledge of the source code and semantic rules to capture the structure and the behavior of design patterns. A tool named Sempatrec (SEMantic PATtern RECovery) has been developed as a plug-in for the Eclipse IDE to implement the approach. Sempatrec processes the Java bytecode of the target software, generates an RDF (Resource Description Framework) ontology and stores the ontology locally in a pool. The reported precision and recall were 61–82% and 88–90% respectively. The accuracy of MLDA has been compared to four approaches as presented in Table 6.

The selection of these approaches was made based on their results, which were detailed enough to compare, and were applied to the same case studies (JHotDraw version 5.1 and JUnit version 3.7). However, the comparison among design pattern detection approaches is challenging. This was due to the fact that there is no standard benchmark to validate the results of each approach. In fact, each approach has its limitations, patterns representation, case studies and validation method. Table 6 shows the results of design patterns extraction of MLDA, Sub-patterns [11], Sempatrec [14], DeMIMA [10] and SSA [9]. As Table 6 illustrates, MLDA achieves reasonable detection accuracy in terms of precision for the detection of JHotDraw and JUnit instances. In addition, MLDA enhances the detection accuracy which relies on the principle of relationships matching. The ability of MLDA to build the structure of each design pattern, record all the object interactions and match the method signatures increase the number of true positive instances. However, MLDA missed the instances that are partly implemented in the source code, since the SSM relies on the standard definition of GoF. On the other hand, the lack of dynamic information explains the existence of false positives. It must be noted that we only compare the results that DeMIMA, SSA, Sempatrec and Sub-patterns revealed.

5 Threats to Validity

Threats to internal validity concern factors that could affect the results. In this paper, this is mainly due to the variants of design patterns. Design pattern instances are recovered based on the standard structural format presented by GoF [1]. Moreover, the way in which the results are validated could affect precision and recall.

Table 6 Comparison of the results of MLDA and that of other approaches

DPs	SS	MLDA			DeMIMA			SSA			Sempatrec			Sub-patterns		
		P%	R%	F%	P%	R%	F%	P%	R%	F%	P%	R%	F%	P%	R%	F%
AD	JD	85	100	92	4	100	8	44	100	61	45			100		
	JU	100	45	63	0			17	100	29	100	100	100	100	100	100
DE	JD	100	33	50	8	100	15	33	33	33	50	33	40	100		
	JU	100	100	100	100	100	100	100	100	100	100	100	100	100	100	100
CO	JD	89	89	89	33	100	50	100	100	100	100	100	100	100		
	JU	NA	NA	NA	100	100	100	100	100	100	100	100	100	100	100	100
FM	JD	NA	0	NA	2	100	4	100	67	80	100	100	100	100		
	JU	100	100	100		100								100	100	100
SI	JD	100	100	100	100	100	100	100	100	100	100	100	100	100		
	JU	NA	NA	NA										100	100	100
OB	JD	NA	0	NA	25	100	40	50	40	44	50	40	44	100		
	JU	NA	0	NA	25	100	40	100	100	100	100	100	100	100	50	67
TM	JD	80	100	89	7	100	13	20	100	33	50	100	67	100		
	JU	50	100	67	0			100	100	100	100	100	100	100	100	100
VI	JD	100	50	67		100		100	100	100				100		
	JU	NA	NA	NA		100									100	100
Average %		90	63	82	34	100	47	74	88	75	83	89	87	100	94	96

Note AD Adapter *DE* Decorator *CO* Command *FM* Factory Method *SI* Singleton *OB* Observer *TM* Template Method *VI* Visitor *SS* Subject Systems *JD* JHotDraw *JU* JUnit *P* Precision *R* Recall *F* F-measure *Blank* Not revealed

To validate the number of true positives, false positives and false negatives, we refer to all publicly published results in the available literature. In fact, we investigated the results of [7, 9–11, 15, 16]. In addition, we used P-MARt [17], the design pattern detection tools benchmark platform [18] and the repository of Perceron [19] as the main benchmarks to validate our results. In doing this, more accurate validation is performed. Threats to external validity concern the generalization of the results. In fact, this paper focuses on Java programming language. It could be worthwhile to conduct the evaluation on other projects having different languages.

6 Conclusion

This paper presented a Multiple Levels Detection Approach (MLDA) to extract design pattern instances from Java source code. MLDA works on three levels: a parsing level, a searching level and a method signatures matching level. The parsing level aims to generate a source code model, which records all objects, classes and methods interaction of the system under study. Furthermore, the parsing level generates a library of design patterns that has the form of source class, destination class and relation type for all GoF design patterns. On the other hand, the searching level introduces the so-called structural search model (SSM), which involves a searching algorithm for each design pattern. The searching algorithm tries to build the pattern structure incrementally based on the generated source code model. The third level of MLDA uses a CLIPS inference engine to match the method signatures of the candidate design instances to that of the subject system. As the experiment results show, MLDA is able to extract 23 design patterns with reasonable detection accuracy.

References

1. Gamma, E., Helm, R., Johnson, R., Vlissides, J.: Design Patterns: Elements of Reusable Object-Oriented Software. Addison-Wesley Longman Publishing Co., Inc., Boston (1995)
2. CLIPS: A Tool for Building Expert Systems. http://www.clipsrules.net/ (2016). 5 Jan 2017
3. Frost, R., Hafiz, R., Callaghan, P.: Parser combinators for ambiguous left-recursive grammars. In: 10th International Symposium on Practical Aspects of Declarative Languages (PADL), ACM-SIGPLAN, vol. 4902, pp. 167–181 (2008)
4. Github. : JavaParser by javaparser. https://javaparser.github.io/javaparser/ (2015). 1 Mar 2015
5. Forgy, C.L.: Rete: a fast algorithm for the many pattern/many object pattern match problem. Artif. Intell. **19**(1), 17–37 (1982)
6. Frakes, W.B., Baeza-Yates, R.: Information Retrieval: Data Structure and Algorithms. Prentice Hall (1992)
7. Guéhéneuc, Y.G., Sahraoui, H., Zaidi, F.: Fingerprinting design patterns. In Proceedings of the 11th Working Conference on Reverse Engineering (WCRE), pp. 172–181. IEEE Computer Society Press, Washington, DC, USA (2004)
8. Guéhéneuc, Y.G., Jussien, N.: Using explanations for design patterns identification. In: Proceedings of the First IJCAI Workshop Modelling and Solving Problems with Constraints, pp. 57–64 (2001)
9. Tsantalis, N., Chatzigeorgiou, A., Stephanides, G., Halkidis, S.: Design pattern detection using similarity scoring. IEEE Trans. Softw. Eng. **32**, 11 (2006)
10. Guéhéneuc, Y., Antoniol, G.: DeMIMA: a multilayered approach for design pattern identification. IEEE Trans. Softw. Eng. **34** (2008)
11. Yu, D., Zhang, Y., Chen, Z. : A comprehensive approach to the recovery of design pattern instances based on sub-patterns and method signatures. J. Syst. Softw. **103**, 1–16 (2015)
12. Shi, N., Olsson, R.: Reverse engineering of design patterns from java source code. In: ASE 06: Proceedings of the 21st IEEE International Conference on Automated Software Engineering, pp. 123–134 (2006)
13. Uchiyama, S., Kubo, A., Washizaki, H., Fukazawa, Y.: Detecting design patterns in object-oriented program source code by using metrics and machine learning. J. Softw. Eng. Appl. **7**, 983–998 (2014)
14. Alnusair, A., Zhao, T., Yan, G.: Rule based detection of design patterns in program code. Int. J. Softw. Tools Technol. Trans. **16**(3), 315–334 (2014)
15. Lucia, A.D., Deufemia, V., Gravino, C., Risi, M.: Design pattern recovery through visual language parsing and source code analysis. J. Syst.Softw. **82**, 1177–1193 (2009)
16. Zanoni, M.: Data mining techniques for design pattern detection. Ph.D. Dissertation, Universita degli Studi di Milano-Bicocca (2012)
17. Guéhéneuc, Y.-G.: P-MARt: Pattern-like micro architecture repository. In: Proceedings of the 1st EuroPLoP Focus Group on Pattern Repositories (2007)
18. Arcelli Fontana, F., Caracciolo, A., Zanoni, M.: DPB: a benchmark for design pattern detection tools. In Proceedings of the 16th European Conference on Software Maintenance and Reengineering (CSMR 12) pp. 235–244. IEEE Computer Society, Szeged, Hungary (2012)
19. Ampatzoglou, A., Michou, O., Stamelos, I.: Building and mining a repository of design pattern instances: practical and research benefits. Entertain. Comput. **4**, 131–142 (2013)

DRSS: Distributed RDF SPARQL Streaming

Amadou Fall Dia, Zakia Kazi-Aoul, Aliou Boly and Elisabeth Métais

Abstract In this work, we present DRSS, a distributed and scalable engine for RDF streams processing. DRSS proposes a new query syntax for continuous querying of RDF data streams. The system includes among others three efficient algorithms for (1) rewriting continuous queries sharing common sub-structures (2), SPARQL query partitioning across multiple computer nodes according to an efficient distribution strategy and (3) query-based data distribution for local processing of sub-queries minimizing data exchanged across nodes. Our system combines both real-time data from multiple sources and stored RDF processing. DRSS and its all algorithms are implemented using the real-time data processing platform Storm Framework, which provides parallelization mechanisms of query operators. The DRSS evaluation is conducted on a real dataset containing up to 1 million RDF graphs. Experiments and obtained results confirm the scalability and the effectiveness of our system.

Keywords DRSS · RDF graphs streams · Distributed sparql

1 Introduction

Data streams are becoming more and more common in many applications like web logs activity, social networking, weather forecast, sensor networks, traffic management, real-time geolocation and so on. Processing such data in near real-time has

A.F. Dia (✉) · Z. Kazi-Aoul
LISITE Lab, ISEP, 75006 Paris, France
e-mail: amadou.dia@isep.fr

Z. Kazi-Aoul
e-mail: zakia.kazi@isep.fr

A. Boly
LID Lab, UCAD, Dakar-Fann, Senegal
e-mail: aliou.boly@ucad.edu.sn

E. Métais
CEDRIC Lab, CNAM, 75003 Paris, France
e-mail: elisabeth.metais@cnam.fr

© Springer International Publishing AG 2018
R. Lee (ed.), *Software Engineering Research, Management and Applications*,
Studies in Computational Intelligence 722, DOI 10.1007/978-3-319-61388-8_8

been widely studied in database community and then, several Data Streams Management Systems (DSMSs) [5] and Complex Event Processing (CEP) systems [11] have been proposed. DSMSs and CEP allow on-the-fly processing of data streams by exploiting the power of continuous query languages such as CQL [4]. However, due to the heterogeneous nature of incoming data and its lack of proper metadata, such systems lack of explicit operators that enable continuous and coherent querying and reasoning over these data from multiple sources. Several work groups have attempted to fill this gap by lifting data streams to semantic level.

The Resource Description Framework (RDF[1]) is the main element for describing data in the semantic web. Data are encoded in RDF model and queried with the SPARQL Protocol and RDF Query Language (SPARQL[2]). The W3C work group RSP[3] (RDF Stream Processing) has extended SPARQL for continuous RDF streaming. The main non-distributed proposed extensions are Streaming SPARQL [8], C-SPARQL [6], $SPARQL_{stream}$ [10], CQELS [17], EP-SPARQL [2] and Sparkwave [16]. Authors have developed a set of operators to support continuous evaluations of SPARQL queries. All the proposed systems provide, among others, solutions to address the heterogeneity issue of data streams but not the parallelism and scalability aspect in RDF stream processing. Indeed, as mentioned in [18, 19], the existing approaches start to fail in many cases such as high number of concurrent queries, large static data or wide range of data generation frequencies. Thus, several approaches such as DIONYSUS [12], CQELS Cloud [19] and C-SPARQL on S4 [15] are proposed for real-time processing of large RDF data in parallel and distributed way. The distributed and highly scalable computing of large RDF streams leads to an efficient data and query partitioning across multiple computer nodes while minimizing data communications between them. Existing systems have proposed advanced techniques for supporting scalability but none of them provides at the same time scalability, combining real-time processing and background knowledge and a full implemented and open source system.

In this work, we introduce DRSS (Distributed RDF SPARQL Streaming), a distributed system for continuous processing of large RDF streams. DRSS addresses all above issues by presenting algorithms (1) for rewriting continuous queries sharing common sub-structures, for (2) SPARQL query partitioning across computer nodes and (3) knowledge-based graph partitioning for locally processing RDF data as much as possible while minimizing data exchanged between machines. Query partitioning and optimization are done "offline" while input data are dynamically distributed over nodes. We also implement a key-value memory storage mechanism of static RDF data to alleviate the cost of join operations between streams and stored graphs. Our system includes both real-time and stored RDF data processing. DRSS is implemented using the real-time distributed platform Apache Storm which provides a set of primitives for parallelization.

---

[1] https://www.w3.org/RDF/.

[2] https://www.w3.org/TR/sparql11-query/.

[3] https://www.w3.org/community/rsp/.

The remainder of this paper is organized as follows. Section 2 is devoted to the presentation of related work in RDF stream processing. Section 3 presents the proposed DRSS architecture and incorporated algorithms. Section 4 provides the details of the implementation before stating the experimental setup and evaluation results of the implemented DRSS system. Finally, Sect. 5 concludes this work and draws some perspectives.

2   Related Work

The task of managing RDF data streams is a well studied problem with many works that mainly extend the semantic query language SPARQL. In this section, we firstly present previous approaches that not offer support for parallel and distributed processing before analyzing systems around distributed computing RDF streams.

2.1   Centralized RDF Stream Processing

Existing RSP languages and associated systems are C-SPARQL [6], SPARQL$_{stream}$ [10], Sparkwave [16] and CQELS [17]. They address a variety of issues including continuous SPARQL query processing and stream reasoning. C-SPARQL and CQELS reuse existing and tested technologies for stream management and knowledge discovering. For instance, C-SPARQL uses ESPER[4] or STREAM [3] and Jena[5] or RDF4J.[6] CQELS implements operators of Aurora [1] and the underlying TDB[7] libraries of Jena. RSP systems use triple-based model where each input event or fact is encoded within one or a sequence of separated RDF triples. However, in the context of `stream reasoning`, events are frequently captured by a set of triples grouped around a RDF graph but not by a succession of separated input triples. Unlike RSP systems, the CEP[8]-based system EP-SPARQL [2] provides an unified language for event processing. The system differs in its way of representing time: temporal operators expressed in terms of sequence and simultaneity. Similar to RSP engines, EP-SPARQL uses RDF triple model as input format. Then, all the approaches used in these systems can not be adopted, for the graph-based model representation (raising the locks of variety and heterogeneity of the data sources) or for the complex and distributed processing of large RDF graphs streams.

---

[4] http://www.espertech.com/esper/.

[5] https://www.jena.apache.org/.

[6] http://www.rdf4j.org/sesame/.

[7] https://www.jena.apache.org/documentation/tdb/.

[8] Complex Event Processing.

2.2 Distributed Stored RDF Data Processing

The problem of distributing RDF graphs into a set of clusters is widely discussed with stored RDF graphs [9, 13, 14, 20, 22, 23]. The distribution process is performed both on the RDF graphs storage and on the SPARQL query execution. The first one is usually performed only once, and a SPARQL 1.1 federated query can be used for executing distributed queries over different SPARQL endpoints. However, this technique is not applicable in streaming context where, new sources and new data are dynamically added at fast and variable speeds. The distribution strategies of RDF graphs streams must be adapted to sliding windows which open the door for the combination of RSP systems to distributed approaches over stored RDF graphs.

2.3 Distributed RDF Stream Processing

In order to efficiently process the large distributed amount of RDF graphs streams, a number of works have been proposed in Semantic Web Community on extending SPARQL around distributed streaming approaches.

- DIONYSUS [12]. Authors introduce a design of a promising system which provides one query interface to enable analytical, streaming and sequence-based queries over distributed RDF graphs. In order to handle data sources variety and heterogeneity issues, authors firstly modeled data as RDF graphs and based their envisioned design on the following four main goals. The first goal is the definition of a Common Basic Graph Pattern store (CBGP-store) which is assigned to a generic BGP and manually or automatically generated from a domain ontology. This storage model allow incremental and indexed computation over sliding windows for an efficient analysis of data at the end. Their second goal is to locally optimize and process an Exact Query Graphs (EQGs) on its foreseen CBGP-store. An EQGs is the query oriented version of a the CBGP-store (i.e. more selective form) with distributed triples patterns and SPARQL1.1 operators (`select`, `optional`, `union`, `filter`, `group by`, `etc.`). Their third goal is to enable different kinds of queries such as analytical, streaming and sequence-based through a single query interface. The fourth and last goal is providing the semantic completeness and locations transparency. This means adding coherence (for instance, new data source added) and query optimization strategies and reduce network traffic by employing local optimization and computation strategies.
- CQELS Cloud [19]. CQELS authors have recently introduced a new system called CQELS cloud. The proposed system is among the first systems addressing elastic and scalable processing for Linked Stream Data. The system distributes the computing and is based on an elastic execution model and a parallelizing algorithms for incremental computing of continuous query operators.

- Elastic execution model for computing continuous queries over RDF streams: the execution model of CQELS Cloud handles a set of continuous query in CQELS query language syntax (CQELS-QL [17]) to produce a set of stream results in one of SPARQL result format. An execution coordinator maps operators among processing nodes. This execution model minimizes communication cost by first using the dictionary encoding approach of CQELS for compression and second deploying on the same machine, operators that consume the same input data.
- Parallelizing algorithms for incremental processing of continuous query operators: each continuous operator of the CQELS-QL (e.g. Aggregation, JOIN and Filter) is considered as complex because of the set of mapping and intermediate results they needed to compute a query.

Each Operator Container (OC) hosts a set of physical query operators that process input streams and forward the output to the consuming operators in the network. The Execution Coordinator coordinates the cluster of OCs using coordination services provided by Storm and HBase[9] which share the same Zookeeper[10] cluster.

The system evaluations[11] show how the throughput scales linearly with the number of processing nodes using LSBench scenario [18]. However, the system only considers matching, join, and aggregate operators and do not include any reasoning task.

- C-SPARQL on S4 [15]. Authors present a new implementation of C-SPARQL query based on the distributed streaming platform S4.[12] The system operates over a fixed RDF schema and implements a partial RDFS reasoning to provide an efficient pattern matching over RDF streams, adding complexity to the system. Authors distributed continuous tasks over set of nodes by considering two subtasks:

  - Closure computation: for distributing triples, authors present a new improved schema which reduces the number of times each triple needs to be processed.
  - Query answering: the system provides a set of stream operators used to support key features in C-SPARQL.

Authors implement the projection, join, selection, filtering and aggregation operators. The system supports time based window and does not implement count-based window. Their implementation also do not support complex time join and can not response to continuous queries with time event overlapping.

---

[9]https://www.hbase.apache.org/.

[10]https://www.zookeeper.apache.org/.

[11]https://www.code.google.com/p/cqels/wiki/CQELSCloud.

[12]http://www.incubator.apache.org/s4/.

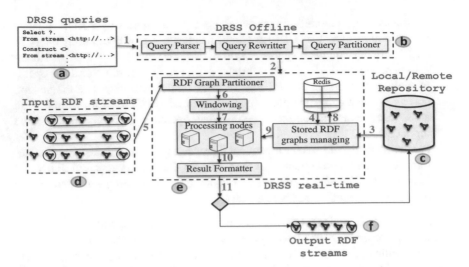

Fig. 1 DRSS system architecture

3 DRSS

In this section, we present our proposed system DRSS (Distributed RDF SPARQL Streaming) and extended SPARQL query language CRSS (Continuous RDF SPARQL Streaming) for continuous querying of RDF graphs within the DRSS system.

Our proposed architecture for Distributed RDF SPARQL Streaming is shown in Fig. 1. This architecture first links six components spread from a → *to* f.

The component a stores all the user queries expressed using the CRSS query syntax (that will be detailed in the following) and is linked to the DRSS offline (b). The Query Parser module splits the CRSS query into dynamic and static part. The dynamic part contents all the identifiers of each stream source within the original query and the window parameters. The static part represents the related SPARQL which is transmitted to the Query Rewritter module and then to the Query Partitioner. These two modules will be explained in detail in the following.

3.1 Query Rewritter Module

The Query Rewritter uses the algorithm 1 for rewriting all SPARQL queries sharing sub-structure(s). A common sub-structure(s) can be designed by one or more triple patterns (located in the WHERE clause). Listings 1 and 2 give two examples of simple SPARQL queries retrieving all stored Ids and geographical location of each sensor observing water pressure in a drinking water distribution network (dwdn).

```
@prefix ex: <http://wdn.org/sensor/el/> .
@prefix ms: <http://datacollect.rsp.org/> .
 SELECT ?sensorId ?lat ?long
   WHERE {
     ?sensId ms:observation "pressure";
          ex:hasLocation ?locatId  .
     ?locatId ms:latitude  ?lat ;
          ms:longitude  ?long .}
```

Listing 1 First query

```
@prefix ex: <http://wdn.org/sensor/el/> .
@prefix ms: <http://datacollect.rsp.org/> .
 SELECT ?sensorId ?sectId ?cityId
   WHERE {
     ?sensId ms:observation "pressure";
          ex:hasLocation ?locatId  .
     ?locatId ms:sectorId  ?sectId;
          ms:city  ?cityId.}
```

Listing 2 Second query

Using the algorithm 1, we notice that the two queries share a triple pattern. The listing 3 shows the rewritten query from both above queries.

```
@prefix ex: <http://wdn.org/sensor/el/> .
@prefix ms: <http://datacollect.rsp.org/> .
 SELECT ?sensorId ?lat ?long ?sectId ?cityId
   WHERE {
     ?sensId ms:observation "pressure";
          ex:hasLocation ?locatId  .
OPTIONAL {?locatId ms:latitude ?latt .}
OPTIONAL {?locatId ms:longitude ?long . }
OPTIONAL {?locatId ms:sectorId ?sectId .}
OPTIONAL {?locatId ms:city ?cityId . }}
```

Listing 3 Rewritten query

Finally, the algorithm 1 regroups all common patterns of various SPARQL queries before rewriting them into a single query. Its advantages are a unique evaluation of common parts to several requests and the non-duplication of a sub query graph process observed in more than one CRSS query. All rewritten queries are then passed to the Query Partitioner.

Algorithm 1: Query Rewritter Module

    Input : Set of SPARQL queries $Q_{initial} = \{Q_{crss_1}, Q_{crss_2}, ..., Q_{crss_n}\}$
    /* Q_{crss_i} are the SPARQL queries extracted from CRSS queries   */
    Output: Set of rewritten SPARQL queries $Q_{Rewritten} = \{Q_1, Q_2, ..., Q_k\}$
    /* with $k \leq n$ ($k = n$ in case of there is no shared
    sub-structure(s) between queries)   */

1 Initialize $S_p = \emptyset$ a set of query graph pattern paths

2 **foreach** $Q \in Q_{initial}$ **do**
3   **if** $Q.formType ==$ *"select"* **then**
4   **Let** *Vars* be the set of variables
5   **Let** *Grp* be the grouping set
6   **Let** *Odr* be the ordering set
7   **end**
8   **Let** *Tp* be the set of triples patterns
9   **Let** *Sp* be the set of sub pattern paths
10   Build the Q graph pattern paths $P_Q = \{Vars, Grp, Odr, Tp, Sp\}$
11    $S_p.\text{add}(P_Q)$
12 **end**

13 Initilialize the set of query path indexer $I_Q = \emptyset$ for each Q
    /* Each query Q has its set of indexer to one or more other
    queries with which it shares a sub-structure   */

14 $i \leftarrow 0$
15 $j \leftarrow 0$
16 **while** $i < S_p.size()$ **do**
17    $j \leftarrow i + 1$
18   **while** $j < S_p.size()$ **do**
19   **if** $P_{Q_i}.sharePatternWith(P_{Q_j})$ **then**
20          $I_{Q_i}.\text{add}(Id_{Q_j})$ /* Id_{Q_j} is the query identifier   */
21
22   **end**
23   **end**
24 **end**

25 **foreach** $p \in P_Q$ **do**
26   **Find** all p shared query indexes for combining them
27    $Q_{path} \leftarrow$ all shared patterns
28    $Q_{optional} \leftarrow$ all not shared patterns
29    $Q_{Rewritten}.\text{add}(Q_{path}, Q_{optional})$
30 **end**

31 **return** $Q_{Rewritten}$

3.2 Query Partitionner Module

Using algorithm 2, the module partitions each SPARQL query based on the join operation. We first list all join nodes present in the query graph pattern and define the notions of `light pattern` (pattern including a single join node) and `full pattern` (formed by exactly two join nodes). `lights` are directly assigned to the corresponding join nodes (and therefore to the corresponding partitions) and the `full` ones are duplicated to their different partitions. This method allows to process a part of the graph pattern locally without requiring communication between the processing nodes. Moreover, only small portions of the graph (`full patterns`) are duplicated between these nodes. Once the different partitions were generated, we used the syntactic and semantic algebra of SPARQL [21] to generate the most optimal query execution plan.

Figures 2 and 3 give respectively an example of the extracted SPARQL query from a CRSS one and its query graph pattern. Figures 4 and 5 show the algebraic join procedure in partitions and all the resulting query graphs partitions. Each partition is processed by one or a set of machines without requiring communications between them. This allows us to parallelize the query process. The resulting query partitions with better execution plans are deployed to the processing nodes (machines) using a distribution and parallelism mechanism based on the Apache Storm topology.

During the chain transmission **b** → *to* **e**, the offline module shares all the information about the query partition Id, the original query Id, the stream Id, the window type, and the window parameters. We need all of them to properly handle and partition the input RDF graphs.

3.3 RDF Graph Partitioner Module

The `DRSS real-time` uses the algorithm 3 for partitioning each incoming RDF graph. The partitioning method remains based on the information gathered from query partitioning step i.e. the query identifier and its partitions, the patterns included in each partition, etc. The interest of this method is to treat each pattern locally.

```
SELECT      ?eventID   ?ObsID ?Result   ?sensorID
WHERE {
            ?eventID   inZone          ?zoneID                              .
            ?eventID   startTime       "2014-01-01T00:00:00.000+01:00" .
            ?eventID   isProducedBy    ?sensorID                        .
            ?eventID   hasValue        ?Result                          .
            ?eventID   hasObservation  ?ObsID                           .
            ?Result    type            ?ObsID                           .
            ?Result    numericValue    1.3E-1                           .
            ?ObsID     inSector        ?sectID                          .
            ?ObsID     type            ObservationValue                 .
            ?ObsID     unit            "CubicMeterPerHour"              . }
```

Fig. 2 A CRSS SPARQL query example

Algorithm 2: Query Partitioner Module

Input : SPARQL query Q, and the partition degree N

Output: Set of SPARQL query partitions $Q_p = \{Q_1, Q_2, ..., Q_n\}$

1   *GraphPatt* $\leftarrow$ *all BGP of Q*
    /* *BGP* $\Leftrightarrow$ SPARQL query basic graph pattern   */

2   Let *joinNode* = {*node* $\in$ *BGP, Degr(node)* > 1 $\wedge$ *node* $\neq$ *predicate*}
    /* *Degr(node)* means the sum of all input and output links of
      the node   */

3   Let *lightPattern* be a semantic link between two nodes where one of
    them is not a *joinNode*

4   Let *fullPattern* be a semantic link between two *joinNode*

5   *NbrOfJoinNodes* $\leftarrow$ *joinNodesCounter(BGP)*

6   *DuplicatePattern* = $\emptyset$

7   $i \leftarrow 0$

8   **while** $i < NbrOfJoinNodes$ **do**

9 | create empty query partition $Q_i = \emptyset$

10 **end**

11 **foreach** *pattern* $\in$ *BGP* **do**

12 | **if** *pattern* is *lightPattern* **then**

13 | | assign *pattern* to the corresponding partition Q_i of the
             current *joinNode*

14 | **else**

15 | | assign *pattern* to both corresponding partitions Q_i and Q_j, $(i \neq j)$
             of the two linked *joinNode*
             /* A pattern shared by two *joinNode* is always duplicated
                 to the two related partitions   */

16 | **end**

17 | **foreach** Partition Q_i **do**

18 | | **if** *currentPartitionDegree(Q_i)* $\geq N$ **then**

19 | | | $Q_p$add(Q_i) /* ignored if Q_i is already in Q_p */

20 | | **end**

21 | **end**

22 **end**

23 **return** Q_p

Before deploying each RDF graph partition to its intended computing node ($\mathbf{d} \rightarrow$ *to* e), we pass it through the windowing systems contained in its original query. Each RDF graph partition is then directed to the windowing module to which it is connected before being redirected to the node intended to process it as soon as it leaves the windowing module. Static RDF data are previously imported (during offline process) and stored into Redis[13] (c $\rightarrow$ *to* e). The queries in the different partitions are formed with the CONSTRUCT header of SPARQL syntax. The intermediate

---

[13]https://www.redis.io/.

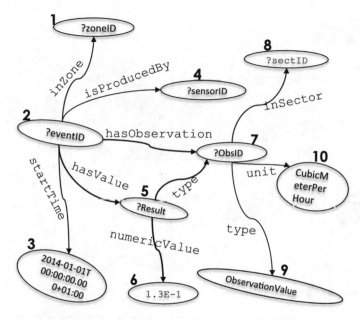

Fig. 3 Corresponding query graph pattern

Fig. 4 Our Algebraic join
procedure

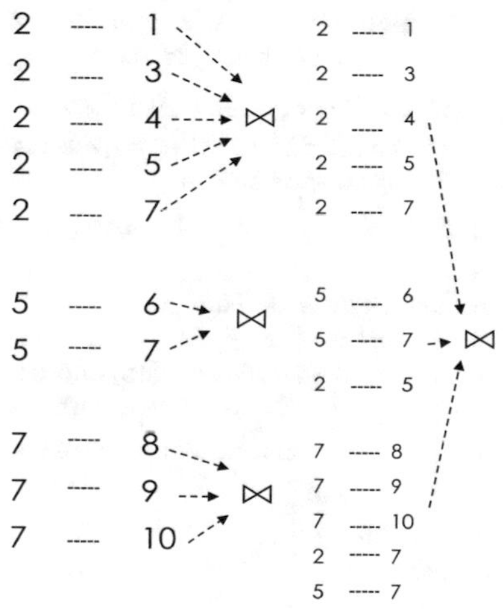

results are supported by the `Result formatter` and, depending on the header
of the CRSS query, will be destined for a local or remote repository or written on an
outgoing stream ($e \rightarrow$ *to* c or $e \rightarrow$ *to* f).

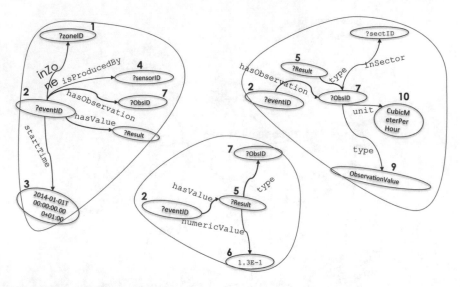

Fig. 5 Resulted query graph partitions

Algorithm 3: RDF Graph Partitioner Module

Input : RDF graph G, set of SPARQL query partitions Q_p
Output: Set of RDF graphs partitions mapped with query
partition id $G_p = \{(G_1, Q_p id), (G_2, Q_p id), ..., (G_n, Q_p id)\}$

1 $M_{QP} = Map < Q_p, List < Pattern >>$
   /* Initialize the map between Query partition id and list of
   used patterns   */
2 $G_{Q_p} = \emptyset$ // the graph associated to a given query partition
3
4 **foreach** *triple* $\in G$ **do**
5 | **foreach** $Q \in Q_p$ **do**
6 | | **if** *queryPatterns*(M_{QP}, Q) *contains triple* **then**
7 | | | $G_{Q_p} \leftarrow$ addNewPattern$(G_Q, triple)$
8 | | | updateGraphSet$(G_{Q_p}, \text{queryId}(Q))$
9 | | **end**
10 | **end**
11 **end**
12 **return** G_p

3.4 CRSS Query Language

We present our continuous extension of SPARQL query named CRSS (Continuous RDF SPARQL Streaming). CRSS is basically a near C-SPARQL [7] query language for continuous processing of RDF graphs streams grouping all the operators of queries present in SPARQL1.1. Thus, in a first step, we have defined a new native query syntax, close to that of C-SPARQL. The presented anatomy of a CRSS query takes into account five (5) variants of windows that we detail through a sample request applied on a case study concerning data from real-time monitoring of a dwdn (pressure, water flow, water level, chlorine level, etc.).

```
QUERY queryID AS {
PREFIX          namespaceIRI
  SELECT | CONSTRUCT | DESCRIBE | ASK
  *FROM [NAMED] STREAM < streamIRI > WINDOW
  *FROM [NAMED]          < staticSourceIRI >
  WHERE { Query graph pattern(s)  }
  GROUP BY expr HAVING expr ORDER BY expr }
  [REFRESH    time  timeUnit]
```

- **[RANGE EVENTS** 1000] returns the last 1000 pressure values observed in a given area.
- **[RANGE EVENTS BATCH** 1000] returns a collection of 1000 pressure values observed in a given area.
- **[RANGE** 10m **STEP** 1m] returns every minute, all pressure values observed in a given area over the last 10 minutes of the implicit timestamp.
- **[EXT RANGE** 10m **STEP** 1m] returns every minute, all pressure values observed in a given area over the last 10 minutes of the explicit timestamp.
- **[RANGE** 10m **EVENT BATCH** 1000] returns the first 1000 pressure values collected in a given sector or those observed after 10 min and which would not have reached the value 1000. This last window is named hybridWind and allows us to prevent peak cases because the first of those satisfied condition (1000 events or 10 min) is immediately processed.

The REFRESH operator is optional and represents the frequency of refreshing the query execution.

4 Implementation and Evaluation

4.1 Datasets Generator and Implementation

In this section, we evaluate our parallelization and distribution strategy over synthetic RDF datasets generated for meeting specific needed complexities that may not be

Table 1 RDF graphs streams generator

Stream Id	LB	UB	Stream rate (G/s)	Topic Id	MAX
w:40fe	20	40	10	fi:40fe	1000000
w:a983	40	60	20	fi:a983	1000000
w:3fef	60	80	30	fi:3fef	1000000
w:9b20	80	100	40	fi:9b20	1000000
w:9eab	100	120	50	fi:9eab	1000000
w:4b4e	120	160	70	fi:4b4e	1000000
w:a1da	160	200	90	fi:a1da	1000000
w:fc3b	200	240	120	fi:fc3b	1000000

found in the available datasets. Datasets relate to the monitoring of a drinking water distribution networks. We use a robust and generic random RDF graphs streams generator which contains a different level of simulation configuration. We give an example of listed characteristics in table I. In the simulation generator, we allocate a temporary `Stream Id` for each stream. This identifier is used for only dissociating the different generated streams and will be replaced by the specified kafka[14] producer `Topic Id` while entered to the kafka queue. Highlight that, in DRSS, each given stream items is written on a unique kafka topic; then, we have the same number of topics as streams.

The simulation runs a parallel workflow (precisely a number of threads equivalent to the number of streams) and then sends all generated streams to their specific kafka topics. A workflow is composed by the graph generation and the graph writing to kafka topic. The graph generation randomly generates an RDF graph i.e., a graph with a number of triples chosen between LB (Lower Bound) and UB (Upper Bound). To avoid false rate caused by the offset observed in graph building, we implement a `warmer` which generates a set of RDF graphs before start sending the graphs. In this case, we make sure to avoid a more longer waiting time than the sleep one needed. The `warmer` time is calculated as follows: Let n be a random value between LB and UB and λ be the time needed to build the graph with $(\frac{LB+n}{2})$ triples. We evaluate the number of optimal graphs N to be generated before start writing them on topic through the following formula:

$$N = n * \lambda.$$

This random RDF graph construct is required to vary the different processing patterns. We need to control incoming data by varying the stream `Rate` in terms of number of RDF graphs per second (`G/s`). Finally, the maximal number of RDF graphs to be generated by each stream is given by the `MAX` parameter (Table 1).

We implement our system using Apache Zookeeper, Apache kafka, Apache Storm[15] and Redis. An input RDF graph is partitioned and each partition is routed

---

[14]https://www.kafka.apache.org/.

[15]http://www.storm-project.net/.

to the same bolt which shares the "equivalent" partition. To define how RDF graphs should be partitioned among the bolts, we use two of eight built-in stream groupings in Storm namely `fields grouping`, `all grouping`. The `fields grouping` is used for grouping all RDF graphs partitions sharing the same query partition Id within the same processing task (processing node). The `all grouping` is used for the replication of a `full pattern` for local and parallel join without communication between nodes. After partitioning, we adopt RDSZ [?] algorithm for RDF stream compression in a distributed environment. RDSZ compresses the items into RDF streams using first a differential encoder followed by Zlib compressor. We first use the URI namespace for reducing the URI names (which are verbose) in the bindings. We then store each created `context` (Pattern+Bindings) in a key-value format into the memory. In this way, all the machines in the cluster will quickly access the `context` for decompression.

4.2   Evaluation Setup and Results

We fully implemented DRSS using Java and all experiments are performed using a cluster of 9 machines running Linux. Each of them has one CPU with 4 cores of 2.4 GHz and 4 GB RAM. Of these 9 machines, there is one we named `fake-node`. The `fake-node` is not a central or master node but is first used for running the `DRSS-offline` queries planner (Query rewriter, Query partitioner and Query optimizer (see Sect. 3) before taking the role of result formatting. The `Result formatter` continuously receives RDF graphs results from computer nodes where only CONSTRUCT sub-queries are executed whenever the header of the registered CRSS query (SELECT, CONSTRUCT or DESCRIBE). Each CRSS query has an identifier which is part of its query partition. The query partition Id is also found on the computing nodes that is intended for processing it. On the real time side, each input RDF graph has an event Id. After partitioning, the new id will be composed by the stream Id, the ids of the queries and partitions that should process this sub-graph. If merging stored RDF graphs with RDF streams is needed, the data importation is done during offline process and stored in Redis. This procedure avoids the risk of blockage in the network during the invocation of remote RDF repositories. The local memory storage allows us to locally join static RDF data to RDF graphs streams in near real-time. It also takes into account a refreshment mechanism of stored data. Indeed, we consider static RDF data as data streams with low frequency refreshment (every 24 h or every week, etc.).

For evaluation, we first consider four DRSS queries **Q1**, **Q2**, **Q3** and **Q4**. For each query, we measure the average execution time considering number of compute nodes and the size of the window. We do not take into account the time needed for result formatting (i.e. integrating sub-results and writing final results on output kafka topic(s) or temporary stored for a future procedure of local or remote repositories refreshment). To test scalability, we vary the size of the cluster and run all 4 queries on clusters with 1, 2, 4, 6 and 8 computing nodes respectively. We normalize

A.F. Dia et al.

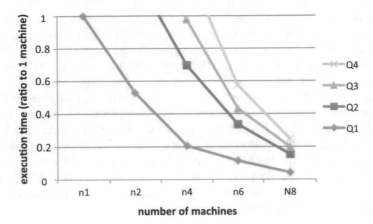

Fig. 6 DRSS scalability test by varying number of computing nodes

the queries execution time to those considered in our baseline where we process all queries on a single node.

In Fig. 6 we observe a low execution time as the number of computing nodes increases. This is due to the local processing of a query partition. There is no shared intermediate result between nodes during the process and the number of join nodes within the query does not have a great influence from 2 machines. This is due to the fact that the join is done in parallel by duplicating the join patterns (`full pattern`).

The above experiments was done with fixed window size. For testing scalability in term of data size to be processed in a window, we resume a similar evaluation by varying the window data size. The results of this evaluation are shown in Fig. 7

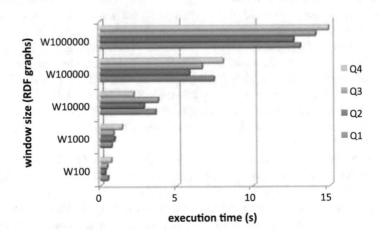

Fig. 7 DRSS sclability test by window length

that plots the evolution of the execution time when we increase the window length in terms of number of RDF graphs. We fixed the input stream rate at 300 graphs/second. Naturally, as the number of RDF graphs to be processed in a window increases, the execution time of each query also increases. The execution time observes a nearly linear evolution confirming the scalability achieved in Fig. 6.

Our last evaluations relate to concurrent queries (**Q1–Q5**) and the combination of multiple streams sources in a query (**Q5**). The given queries are not interdependent but, we configure their outputs such a way that they depend on each others. We first consider a set of concurrent queries (i.e. queries that need results provided by one or more queries in the same topology running) and measure the throughput RDF graphs (RDF graphs/second) while fixing the window size to 100,000 graphs for all queries and varying the number of computing nodes. The links observed between used queries can be represented as follows:

- **Q2** → *needs* **Q1**
- **Q3** → *needs* **Q1 & Q2**
- **Q4** → *needs* **Q1 & Q2 & Q3**
- **Q5** → *needs* **Q1 & Q2 & Q3 & Q4**

In Fig. 8 we tend to have a linear growth of the throughput when we increase the number of computing nodes. However, the throughput is negatively proportional to the number of queries that a query depends. But this evolution can always be offset by additional computing nodes. Given that **Q5** does not use `left outer join` and the possibility that the internal query (query on which depends **Q5**) does not return result the query **Q5** does not return any result during processing time. Similarly, we can do the same observation in Fig. 9 for the variation of the number of source streams joined within the query **Q5**.

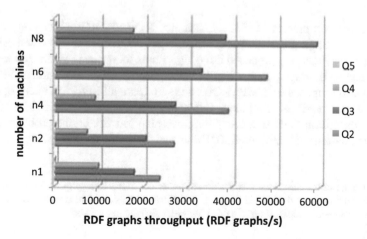

Fig. 8 DRSS concurrent queries evaluation

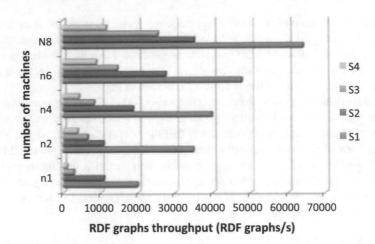

Fig. 9 DRSS multiple streams evaluation

The more computing nodes are added, the more the expected performances believe. These experiments demonstrate the scalability of our parallelization strategy of the processing and the distribution of the data.

Highlight that we do not test the elasticity of our system but we have configured a robust mechanism which allocate an input task (with partition Id) dedicated to a bolt which has not sent acknowledgment (Ack), to an another bolt.

5 Conclusion and Future Work

In this paper, we present DRSS, a distributed RDF SPARQL Streaming for scalable and continuous processing of continuous SPARQL queries expressed using our CRSS query syntax. We proposed three algorithms for query rewriting, query partitioning and RDF graph partitioning. We also propose a new extension of SPARQL neared those proposed by C-SPARQL and regrouping all SPARQL query operators with three new variants of window. Experiments show excellent scalability in term of query execution time. In a near future, we plan to work on the problem of scalable combinations of large stored RDF data and their refreshment during continuous process.

Acknowledgements This work was performed under the FUI Waves project. This project aims to design and develop a distributed processing platform of massive data streams. The case study concerns the real-time monitoring of a drinking water distribution network.

Appendix

QUERIES

- **Q1**: *Query q1 AS {SELECT ?eID ?obsID ?result ?sID FROM STREAM <http : //ex.org/obs> [RANGE EVENTS 5000] WHERE { ?eID zone ?zID . ?eID start-Time "2014-01-01T00:00:00.000+01:00" . ?eID isProducedBy ?sID . ?eID has-Value ?result . ?eID hasObservation ?obsID .?result type ?ObsID . ?result num-Value 1.3e-1 . ?obsID inSector ?sectID ?obsID type observationValue .?obsID unit "CubicMeterPerHour".}}*

- **Q2**: *Query q2 AS { SELECT ?pressureSens ?value FROM STREAM <ex.org/obs> [RANGE 60s STEP 20s] FROM <http : //exorg/staticrepo> WHERE {?sector rdfs:label "Lou" . ?pressureSens ssn:onPlatform ?sector. ?event ssn:isProducedBy ?pressureSens; ssn:hasValue ?observation. ?observation a "pressure" ; w:numValue ?value.} }*

- **Q3**: *Query q3 AS SELECT ?pressureSens ?value FROM STREAM <ex.org/obs> [RANGE 60s STEP 20s] FROM <http : //exorg/staticrepo>WHERE {?sector rdfs:label "Lou" . ?pressureSens ssn:onPlatform ?sector . ?event ssn:isProduced By ?pressureSens; ssn:hasValue ?observation. ?observation a "pressure" ; w:num Value?value.}}*

- **Q4**: *Query q4 AS SELECT ?sensor ?value FROM STREAM <http : //ex.sh> [RANGE EVENTS BATCH 1000] WHERE {?eventID ssn:hasValue observation; ssn:isProducedBy ?sensor ; a ssn:SensorOutput. ?observation qudt:numValue ?value; qudt:unit "http://qudt.org/unit#CubicMeterPerHour"; a ssn:Observation Value .}}*

- **Q5**: *QUERY q5 AS { SELECT ?aSub ?anObj (count(distinct ?subject) as ?count) FROM STREAM < http : //example.org/observations1 > [RANGE EVENTS 10000] FROM NAMED < http : //repository.org/biblio.rdf > FROM STREAM < http : //example.org/observations2 > [RANGE 1m STEP 30s] FROM NAMED < http : //repository.org/book.rdf > WHERE {?subject weather:type ?object . ?aSub sens-obs:aProp ?anObj ; sens-obs:anotherProp ?anotherObj ; { select ?name (count(distinct ?object) as ?count2) FROM STREAM < http : //example.org/observations2 > [RANGE 2m STEP 30s] FROM STREAM < http : //example.org/observations3 > [RANGE 10m TUMBLING] FROM NAMED < http : //repository.org/biblio.rdf > WHERE {?object weather:name ?name ; sens-obs:aProp ?aName .} GROUP BY ?name }} GROUP by ?aSub ?anObj }}*

References

1. Abadi, D.J., Carney, D., Çetintemel, U., Cherniack, M., Convey, C., Lee, S., Stonebraker, M., Tatbul, N., Zdonik, S.: Aurora: a new model and architecture for data stream management. VLDB J. Int. J. Very Large Data Bases **12**(2), 120–139 (2003)
2. Anicic, D., Fodor, P., Rudolph, S., Stojanovic, N.: Ep-sparql: a unified language for event processing and stream reasoning. In: Proceedings of the 20th International Conference on World Wide Web, pp. 635–644. ACM (2011)
3. Arasu, A., Babcock, B., Babu, S., Cieslewicz, J., Datar, M., Ito, K., Motwani, R., Srivastava, U., Widom, J.: Stream: the stanford data stream management system. In: Data Stream Management, pp. 317–336. Springer (2016)
4. Arasu, A., Babu, S., Widom, J.: The cql continuous query language: semantic foundations and query execution. VLDB J. Int. J. Very Large Data Bases **15**(2), 121–142 (2006)
5. Babcock, B., Babu, S., Datar, M., Motwani, R., Widom, J.: Models and issues in data stream systems. In: Proceedings of the Twenty-first ACM SIGMOD-SIGACT-SIGART Symposium on Principles of Database Systems, pp. 1–16. ACM (2002)
6. Barbieri, D.F., Braga, D., Ceri, S., Grossniklaus, M.: An execution environment for c-sparql queries. In: Proceedings of the 13th International Conference on Extending Database Technology, pp. 441–452. ACM (2010)
7. Barbieri, D.F., Braga, D., Ceri, S., Valle, E.D., Grossniklaus, M.: C-sparql: a continuous query language for RDF data streams. Int. J. Semant. Comput. **4**(01), 3–25 (2010)
8. Bolles, A., Grawunder, M., Jacobi, J.: Streaming sparql-extending sparql to process data streams. Semant. Web: Res. Appl. 448–462 (2008)
9. Buil-Aranda, C., Arenas, M., Corcho, O., Polleres, A.: Federating queries in sparql 1.1: syntax, semantics and evaluation. Web Semant.: Sci. Serv. Agents World Wide Web **18**(1), 1–17 (2013)
10. Calbimonte, J.P., Corcho, O., Gray, A.J.: Enabling ontology-based access to streaming data sources. In: The Semantic Web–ISWC 2010, pp. 96–111. Springer (2010)
11. Etzion, O., Niblett, P.: Event Processing in Action. Manning Publications Co. (2010)
12. Gillani, S., Picard, G., Laforest, F.: Dionysus: towards query-aware distributed processing of RDF graph streams. In: EDBT/ICDT Workshops. Citeseer (2016)
13. Gurajada, S., Seufert, S., Miliaraki, I., Theobald, M.: Triad: a distributed shared-nothing rdf engine based on asynchronous message passing. In: Proceedings of the 2014 ACM SIGMOD International Conference on Management of Data, pp. 289–300. ACM (2014)
14. Hammoud, M., Rabbou, D.A., Nouri, R., Beheshti, S.M.R., Sakr, S.: Dream: distributed rdf engine with adaptive query planner and minimal communication. Proc. VLDB Endow. **8**(6), 654–665 (2015)
15. Hoeksema, J., Kotoulas, S.: High-performance distributed stream reasoning using s4. In: Ordring Workshop at ISWC (2011)
16. Komazec, S., Cerri, D., Fensel, D.: Sparkwave: continuous schema-enhanced pattern matching over RDF data streams. In: Proceedings of the 6th ACM International Conference on Distributed Event-Based Systems, pp. 58–68 (2012)
17. Le-Phuoc, D., Dao-Tran, M., Parreira, J.X., Hauswirth, M.: A native and adaptive approach for unified processing of linked streams and linked data. In: The Semantic Web–ISWC 2011, pp. 370–388. Springer (2011)
18. Le-Phuoc, D., Dao-Tran, M., Pham, M.D., Boncz, P., Eiter, T., Fink, M.: Linked stream data processing engines: facts and figures. Semant. Web-ISWC **2012**, 300–312 (2012)
19. Le-Phuoc, D., Quoc, H.N.M., van Le, C., Hauswirth, M.: Elastic and scalable processing of linked stream data in the cloud. In: International Semantic Web Conference, pp. 280–297. Springer (2013)
20. Makris, K., Bikakis, N., Gioldasis, N., Christodoulakis, S.: Sparql-rw: transparent query access over mapped RDF data sources. In: Proceedings of the 15th International Conference on Extending Database Technology, pp. 610–613. ACM (2012)
21. Pérez, J., Arenas, M., Gutierrez, C.: Semantics and complexity of sparql. ACM Trans. Database Syst. (TODS) **34**(3), 16 (2009)

22. Schwarte, A., Haase, P., Hose, K., Schenkel, R., Schmidt, M.: Fedx: a federation layer for distributed query processing on linked open data. In: Extended Semantic Web Conference, pp. 481–486. Springer (2011)
23. Zeng, K., Yang, J., Wang, H., Shao, B., Wang, Z.: A distributed graph engine for web scale rdf data. In: Proceedings of the VLDB Endowment, vol. 6, pp. 265–276. VLDB Endowment (2013)

An Efficient Approach for Real-Time Processing of RDSZ-Based Compressed RDF Streams

Ndéye Bousso Déme, Amadou Fall Dia, Aliou Boly, Zakia Kazi-Aoul
and Raja Chiky

Abstract In recent years, the volume of generated RDF graphs streams from different fields of applications is very large and therefore difficult to process in an optimized manner. Indeed, processing such data in conventional triplestores can be costly in terms of execution time and memory consumption. Several works have examined data compression approach both on static and dynamic RDF data. In addition to those based on stored RDF data, two recent compression algorithms RDSZ and ERI were focused on RDF streams. Continuous compressed format requires less memory space but cannot be exploited through SPARQL queries. In this paper, we propose an approach for continuous querying RDSZ-based RDF streams without decompression phase. We add three algorithms from simple to aggregate query execution over RDSZ compressed items. Our experimentation use real datasets to demonstrate the effectiveness and efficiency of our proposition in term of query execution time and memory save.

Keywords RDF streams · RDSZ · Compression · Continuous querying

N.B. Déme · A. Boly
LID Lab, UCAD, Dakar-Fann, Senegal
e-mail: ndeyebousso.deme@ucad.edu.sn

A. Boly
e-mail: aliou.boly@ucad.edu.sn

A.F. Dia (✉) · Z. Kazi-Aoul · R. Chiky
LISITE Lab, ISEP, 75006 Paris, France
e-mail: amadou.dia@isep.fr

Z. Kazi-Aoul
e-mail: zakia.kazi@isep.fr

R. Chiky
e-mail: raja.chiky@isep.fr

© Springer International Publishing AG 2018
R. Lee (ed.), *Software Engineering Research, Management and Applications*,
Studies in Computational Intelligence 722, DOI 10.1007/978-3-319-61388-8_9

1   Introduction

In recent years, large volumes of data has been continuously generated at very fast speeds by different fields of applications such as social networks, websites logs, sensor networks, geolocation, quantified-self measurement, logistics management, traffic management, alarms triggering, etc. Data are issued from multiple sources and follow different formats (XML, CSV, textual, etc.). Due to the data heterogeneity, the semantic web community, with its broad spectrum of technologies (RDF, SPARQL, RDFS and OWL), has defined a common model for representating and processing heterogeneous data. More recent works propose to use semantic web technologies to process continuous RDF streams by providing RSP technologies (RDF Stream Processing). Among these systems we can cite C-SPARQL [5], CQELS [17], SPARQL$_{stream}$ [7], EP-SPARQL [3] and Sparkwave [16]. They allow real-time processing of RDF triples by providing windowing operators or continuous querying.

However, it can be very difficult to process the input RDF streams due to their huge volume and its velocity. In order to reduce the cost of data processing, prior works have adopted RDF data streams partitioning [1], summarization [8] and compression. RDF data compression has given rise to recent works on static (stored) and dynamic RDF data (streams). Since the "static" approaches [2, 11, 13, 15, 18] are not suitable to RDF streams, two recent solutions RDSZ [14] and ERI [12] proposed algorithms for RDF data streams compression. Unfortunately, data issued form RDSZ or ERI compression algorithms cannot be processed (SPARQL queries execution) without prior decompression phase. Decompression phase naturally increases the processing cost in term of execution time and memory consumption.

In this paper, we propose an extension of the RDSZ algorithm allowing to query compressed data immediately. We propose three equivalent algorithms for processing simple SPARQL query, SPARQL queries with FILTER operator(s) and SPARQL queries with aggregate operator(s) over RDSZ-based compressed RDF data. We evaluated our approaches using a real world RDF dataset that contains expressive sensors description of 20,000 weather stations.

The remainder of this paper is organized as follows: Sect. 3 presents the existing solutions on RDF data compression. We present our approach in Sect. 4 which we will evaluate in Sect. 5. Section 6 presents the conclusion and some perspectives of our work.

2   Background and Preliminaries

Integration, processing and analysis of data from different sources require common representation model. The semantic web was proposed by Tim Berners-Lee (the inventor of the web). The term refers to an evolution of the syntactic web. According to Tim Berners-Lee, its implementation would allow automation of a large number

Fig. 1 Semantic web stack

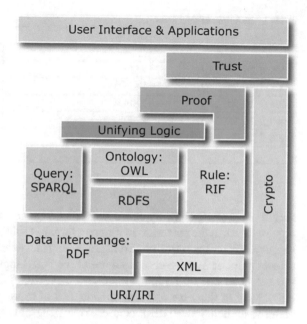

of tasks and easier access to knowledge. From a technological point of view, the semantic web is represented by a broad spectrum of technologies making up its theory. We present in Fig. 1,[1] a widely distributed schema, better known as "the semantic web layer Cake", and which summarizes the semantic web standard stack, thus materializing the roadmap drawn up by Tim Berners-Lee [6]. The ascending stack representation materializes the dependencies between the different standards (each layer exploiting the previous layers). In the following, we present only the tools we used in our work and those needed to understand them.

RDF (Resource Description Framework)

RDF is the lowest layer among semantic web-specific standards. It is a labeled graph template for describing resources formally. It exploits the URI formalism and imposes a form of subject—predicate—object triples for the expression of knowledge. An RDF triple can be represented as a graph. An RDF Graph is a set of triples where subjects and objects represent the nodes and predicates form the links between nodes.

SPARQL (SPARQL Protocol and RDF Query Language)

SPARQL is the W3C standard for querying **RDF** triples. It is both a language and a query protocol. The protocol will allow a Web client to consult a service or SPARQL

---

[1]https://www.w3.org/2007/03/layerCake.png.

Table 1 Example of data stream

Timestamp	SensorID	Wind speed
...	...	...
2008_8_26_21_24_00	ZSFO1	"41.0"
2008_8_26_21_24_03	ZSFO1	"40.8"
2008_8_26_21_24_06	ZSFO1	"40.9"
2008_8_26_21_24_09	ZSFO1	"41.0"
2008_8_26_21_24_12	ZSFO1	"41.2"
...	...	...

access point (endpoint), by executing a SPARQL request, which will process the request to return the responses in different formats (RDF/XML, N3, JSON, etc.). The language allows querying, modifying, inserting and deleting RDF descriptions using clauses (similar in some cases to those of the SQL language). SPARQL query head form must be one of the following clauses: SELECT, CONSTRUCT, DESCRIBE, ASK. Most of these query forms contain a set of triple patterns called a basic graph pattern and included WHERE clause. Triple patterns are like RDF triples except that each subject, predicate, and object can be a variable.

Today, many systems are continuously receiving large amounts of data generated at high speeds. A Data stream is defined in [9] as an infinite sequence of continuously generated elements at a fast rate (Table 1).

Windowing in Data Streams

One of the peculiarities of the analysis of data streams with respect to that of the conventional data is its temporal aspect. Moreover, since a data stream is infinite, it cannot be treated in its whole entirety. It is therefore necessary to define the portion of the stream to which a treatment will be applied. To do this, we use windows that mark the portion of the stream to be processed by referencing a start date and an end date. It is thus possible to request only a sample of streams and not the entire stream, accelerating the requests since the streams are continuous elements and therefore without size limits. There are mainly four window types: fixed window, benchmark window, sliding window and logical window. The first three are physical windows based especially on the temporal aspect, and the last one constitutes a windowing based on the number of elements in the stream.

Physical Window

It's used to create windows that mark the portion of the stream to be processed by means of a start date and an end date. There are three types of physical windows detailed in the following.

Fixed Window

is a window where the start and end dates are absolute. For example, a window between the January 1st, 2016 at 11:00:00 PM and March 31st, 2016 at 11:00:00 PM.

Benchmark Window

is a window where one of the dates is relative, the reference point window is called. For example, a window between December 31st, 2010 at 00:00:00 and the current date.

Scrolling or Sliding Window

This is the case if the two dates are relatives. For example, a window covering the last 10 days or hours.

Logical Window

It is also called a sequential window. The dates are expressed in the order of arrival of the elements of the stream. For example, a window on the last ten elements of the stream or in a more flexible way, a window on the 1000*th* and the 2000*th* elements of the stream.

Data Stream Management System (DSMS)

Relational databases remain ubiquitous in the enterprise where applications require constant storage of data and queries ranging from the simplest to the most complex. Data are stored statically and update are performed less frequently than tables and join selection queries. Queries are executed when requested by users and the provided result reflects the consistent state of the database. These databases are under the implementation of relational database management systems (RDBMS). New applications are emerging and they require a new type of data storage and query. Today, processing time becomes increasingly critical for some applications. They need to query and analyze data faster, in order to make decisions or react as quickly as possible. From the control of network traffic to the analysis of transactional logs (web, banking or telecommunication transactions) and the management of sensors data, most applications require data management systems able to manage continuous data. Taking into account criteria that differ from conventional data. Data management systems (DSMSs) have been designed to meet the needs of many applications for continuous process of generated data. Unlike DBMSs, DSMSs work with transient and persistent queries.

Semantic Data Streams

The extraction of knowledge on classical (static) using semantics, has proved its worth in many fields of application. Moreover, with the recurring problem of "large volume of data processing and knowledge extraction", existing techniques are not sufficient to obtain and derive information from the large amount of data available. It is necessary to combine the techniques of representation of the static data in RDF format with the characteristics and the information contained in the streams in order to lead to the reasoning and the extraction of pertinent information. Therefore, a new concept called RDF Stream has been defined in [4, 10], etc.

An **RDF stream** is a new data type, and is an RDF extension for the semantic representation of data stream. RDF Stream is an extension to the RDF format by adding a timestamp. Thus, a RDF Stream is defined as an ordered sequence of pairs, where each pair is formed of an RDF triple and its timestamp τ.

$$(< subject_1 \ predicate_1 \ object_1 >, \tau_1)$$
$$(< subject_2 \ predicate_2 \ object_2 >, \tau_2)$$
$$...$$
$$(< subject_i \ predicate_i \ object_i >, \tau_i)$$

The timestamp can be considered as an annotation of the RDF triples, and is monotonically non-decreasing between the streams. Specifically, the timestamp is not strictly increasing because it is not necessarily unique. Two or more consecutive triples may have the same timestamp, meaning that they occur at the same time.

3 Related Work

In this section, we present the existing RDF data compression techniques.

3.1 Approaches over Stored RDF

Several works have been performed for the compression of static RDF data.

3.1.1 Basic Approaches

In [11], several approaches are proposed to compress the RDF data. The first are using compressors gzip, bzip2 and ppmdi to compress RDF data. The second approach uses an adjacency list. The third and final approach use an encoding dictionary. The conclusions drawn by the authors are that RDF data are largely compressible and that conventional compression techniques are not applicable on RDF data.

3.1.2 Logical Linked Data Compression

In [15], authors propose an algorithm that compresses real world datasets by generating a set of logical rules from the dataset. The triples deducted from these rules are then deleted. Indeed, the algorithm automatically generates a set of rules and divides the database into two sets of smaller disjoint data, namely a set of active data and an inactive data set based on these rules. The inactive dataset contains a list of uncompressed triples that remain and to which no rule can be applied during decompression phase. On the other hand, the active data set contains a list of compressed triples, to which rules are applied. The evaluation of this algorithm shows that more than half of the triples can be removed without any loss of integrity.

3.1.3 Scalable RDF Data Compression with MapReduce

The RDF data compression technique with MapReduce proposed in [18] allows to compress and decompress large amount of RDF data in a parallel way. They use a dictionary coding technique that maintains the data structure and each term in a dataset is replaced by a numeric ID.

3.1.4 Compressed k2-Triples for Full-In-Memory RDF Engines

The k2-triples algorithm in [2] uses a compact indexed RDF structure (called k2-triples) applying compact k2-tree structures to the well-known vertical partitioning technique. The result is an ultra-compressed representation of large RDF graphs and allows the SPARQL queries to be executed in memory without decompression.

3.1.5 HDT (Header, Dictionary, Triples)

HDT in [13] is a compact data structure and a binary serialization format for RDF. HDT can compress large datasets to save space while maintaining search and navigation without pre-decompression. HDT is a representation format based on 3 components:

- A header which contains logical and physical metadata that describes the RDF dataset.
- A dictionary which organizes the identifiers of the RDF graphs.
- A triple set which includes the pure structure of the RDF graphs.

3.2   Approaches over RDF Streams

Two recent works proposed techniques over RDF data streams compression.

3.2.1   RDSZ

The RDSZ (RDF Differential Stream compressor based on Zlib) algorithm in [14] is an RDF stream compression algorithm without loss of data. RDSZ takes advantage of the fact that the items in an RDF stream have structural similarities that can be exploited by a differential encoding mechanism so that the new elements in the stream can be represented on the basis of the previously processed ones. During the compression phase the RDSZ algorithm uses three components:

The **differential encoder** which takes as input a sequence of elements of an RDF stream. The RDF elements at the input are processed sequentially and separately by the encoder. The first processing performed on an element is its decomposition into a pattern of triples and an array of variables and corresponding values called bindings. It returns as a result a string that represents the pattern ($?x1$ <predicate> $?x2$) obtained as output from the replacement process and a table of link variables that binds each variable to its particular value of the input. After obtaining the pattern and the binding, the encoder needs to determine whether this element can be represented based on an element already processed in the stream or not. To do this, the encoder uses the information on previously processed elements that are stored in a cache called LRU (Least Recently Used). For each pattern, the associated variables and a unique identifier are stored. After reading the cache, if the pattern of the RDF element being processed is already in the cache, it means that another element with the same pattern has been recently processed. Thus, the active element is coded on the basis of the previous one. Since the two elements have the same pattern, there is no need to re-send all the pattern data to the decompressor. Only the pattern identifier will be included in the coded element. As for the variables, the new element and the preceding one can have the same values or not. If a variable value is the same, there is no need to send it again. Otherwise, the value is included in the coded element. In this case, the result of the encoding process is a string that contains a row for the model identifier and a row for each variable.

The **multiplexer** which takes as input an element sequence (coded or not) and converts it into a single string by concatenating the serialization text of the elements. A special delimiter is used to mark the boundaries of each element, so that the decompressor can separate them again. The **compressor** which takes as input the string generated by the multiplexer and compresses this string. The Zlib compressor implements the deflate algorithm which is a lossless data compression algorithm that couples the LZ77 algorithm and the Huffman coding.

3.2.2 ERI

The ERI (Efficient RDF Interchange) format in [12] is a compact representation of RDF designed to take advantage of the redundancy of structures and inherent data of RDF streams. ERI is based on RDSZ and uses the fact that in most RDF steams the structure of the information is well known by the data provided and the number of variations in the structure is limited.

3.2.3 Evaluation and Comparison Between RDSZ and ERI

We evaluate both algorithms in order to choose one of them according to some criteria: compression and decompression time, size of data after the compression step and the accuracy of the results of the queries on data after compression–decompression. For the RDSZ algorithm, we use the following configuration: batchSize = 5 (number of elements processed by the compressor) and cacheSize = 100 (cache size). As for ERI, it offers multiple ERI-1K (blocksize = 1024), ERI-4k (blocksize = 4096) and ERI-4k-Nodict (blocksize = 4096) configurations. ERI-1K and ERI-4K contain an LRU dictionary for each value channel, while ERI-4k-Nodict does not contain it. For the evaluation, the ERI-4K and ERI-4k-Nodict configurations were chosen. The datasets used to make the comparison are those of RDSZ[2] (AEMET2, Identica, Wikipedia, Petrol, LOD, Mix).

3.2.4 Size of Data Obtained After Compression

For the size of the data obtained after compression, the two algorithms remain comparable. In some cases ERI surpasses RDSZ because when dividing the datasets into graphs (RDSZ takes as input a graphs stream), elements with similarities are separated. On the other hand, RDSZ compression slightly surpasses ERI in particular cases where the predicate number is small. In such cases, is more costly due to the use of multiple compression channels.

3.2.5 Compression and Decompression Time

ERI compression time is significantly faster than RDSZ. Indeed, RDSZ processes and exits graph streams while ERI processes triplet ones. Thus, RDSZ compression may be affected by the fact that it must potentially process very large graphs with several triplets, whereby the differential coding process takes a longer time. On the other hand, the decompression time RDSZ is faster than ERI because ERI decompresses several channels and spells all the triplets of a block.

---

[2]http://www.it.uc3m.es/berto/RDSZ/.

Table 2 Comparison between RDSZ and ERI

Algorithm	Compression time	Decompression time	Size of data after compression	Accuracy of results
ERI	-	+	+/-	-
RDSZ	+	-	+/-	+

3.2.6 Accuracy of Results

Since RDSZ is a lossless data compression algorithm, when a query is executed after the decompression phase we get the same query as when we run the query before the compression phase. However, when compressing with the ERI algorithm, data loss can occurs. The table below summarizes the comparison between the two algorithms.

In our state of the art, we have seen that there are two approaches for compressing RDF data. The first approach is not adapted to RDF graph stream processing since it is based on static data. In the second, two algorithms have been proposed. However, to execute queries on the compressed data with these two algorithms, it is necessary to go through a decompression step. In our contribution we will extend RDSZ algorithm in order to execute the queries on the compressed data. Our choice is motivated by the Accuracy of results criterion. Indeed, the evaluation and comparison of the two algorithms (Table 2) which enabled us to confirm that, unlike the ERI, RDSZ is a lossless compression algorithm. In the following section we will describe our contribution based on this algorithm.

4   Real-Time Querying over RDSZ-based Compressed RDF

4.1   System Architecture

One solution to reduce the cost of query operation on RDF streams is to execute queries after the compression step. In our contribution, we implement this solution for three forms of queries: **Simple SPARQL query**, **SPARQL Query with filter operator(s)** and **SPARQL Query with aggregate function(s)**.

As described in Fig. 2, the architecture of our system is composed of several modules. The first step for any kind of SPARQL query is the initialization phase. The system takes as input three parameters: the SPARQL query, the window type and the window size. During the initialization phase, several tasks are performed based on these parameters. The first task is to retrieve patterns from the SPARQL query. Indeed, a SPARQL query can consist of several patterns, during each phase, we parse the input query. The patterns are retrieved and put into a list which is used in next

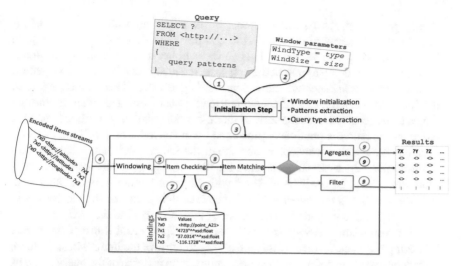

Fig. 2 System architecture

step. The second phase is the initialization of the window with the given input size. The window allows us to define the number of elements to be considered before starting the treatment. The third and last task of this phase is the retrieval of the type of the request. Indeed, the system can handle three types of requests. The first type concerns simple queries. These queries are of the form SELECT ... WHERE {...} without filter or aggregate function. The second type concerns requests with filter operator(s). Two types of filter are managed: the arithmetic filtering SELECT ... WHERE { ... FILTER (...) } and the string expression filtering SELECT ... WHERE { ... FILTER regex(...) }. Arithmetic expression filtering are applied on integers, real numbers, etc. Several types of operators can be applied to these types, such as: superiority, inferiority or equality. For queries with string expression filtering, two cases have been managed: the equality of two strings and the capacity of a sub-string of characters by a string. The last type of request concerns aggregate operators such as COUNT, SUM and AVG. In the following we will detail the processing of these three types of queries.

Simple Query

After the initialization phase of the system, the system waits for the item to be encoded by the RDSZ algorithm. Indeed, the user defines a window that contains the items to be encoded by RDSZ. Therefore, when the window reaches its size, the processing of the items contained in this window begins. We detail this processing in the following.

- **Job 1**: **checking the pattern**: Verifying the item patterns returned by the RDSZ algorithm is an essential step in processing the query. Indeed, during the compression step of the RDSZ algorithm, the items can be encoded based on those already processed. As explained in Sect. 3, the RDSZ algorithm uses the redundancy on the patterns of the datasets to compress the RDF data. Thus, when an item is received, it can be encoded in two ways: either the same pattern was previously processed, or it was not. In the first case, the identifier of the previous pattern is used, while in the second a new identifier is generated. Then, the objective of this job is to check the type of encoding used for each item. The job goes through the patterns received, and for each of them, checks whether it has been encoded based on the patterns previously processed by the encoder or not. If the item has been encoded with a previous pattern, then the pattern and pattern binding are retrieved from the cache. Indeed, since the item has been encoded based on a previous item, the pattern will be retrieved with the identifier of this item. Otherwise, only the binding is recovered in the cache since the pattern of the received item will correspond to its pattern. The output of this job is sent to the next job.

- **Job 2**: **comparison between the patterns of the query and those of the items**: The comparison performed by this job allows to retrieve the items whose patterns correspond to those defined in the query. Indeed, the patterns of the query allow us to make a filter on the items to return after the execution of the query. As described in the Algorithm 1, the job goes through the patterns and binding of each item sent by the previous job by retrieving the subject, predicate, and object of the pattern. Then, it goes through the set of patterns of the query that were recovered in the initialization phase by also retrieving the subject, the predicate and the object of the pattern. Patterns of the query can be presented according the following eight (8) different combinations:

- The three subject, predicate, and object elements are variables (?subject ?predicate ?object)
- The subject is a constant and the predicate and the object are variables (<http://pointA21> ?predicate ?object)
- The subject and the predicate are constants and the object is a variable (<http://pointA21> <http://latitude> ?object)
- The three elements (subject, predicate and object) are constants (<http://pointA21> <http://latitude> "4723")
- The subject and the object are variables and the predicate is a constant (?*subject* <http://latitude> ?object)
- The subject and the predicate are variables and the object is a constant (?subject ?predicate "4723")
- The subject is a variable and the predicate and the object are constants (?subject <http://latitude> "4723")
- The subject and the object are constants and the predicate is a variable (<http://pointA21> ?predicate "4723")

Algorithm 1 Matching item step

```
 1: function MATCHEDITEM(encodedItem, cache)
 2:     result ← null
 3:     foreach item in encodedItem do
 4:         diffmodel ← checkItem(item)
 5:         itemPatterns ← diffmodel.getPattern()
 6:         itemBindings ← diffmodel.getBindings()
 7:         queryPatterns ← query.getPattern

 8:         foreach pattern in queryPatterns do
 9:             queryS ← pattern.getS()
10:             queryP ← pattern.getP()
11:             queryO ← pattern.getO()

12:             if queryS.IsVar && queryP.IsVar && queryO.IsVar then
13:                 foreach itemPattern in itemPatterns do
14:                     itemS ← Bindings.getS(patternQuery)
15:                     itemP ← patternQuery.getP
16:                     itemO ← Bindings.getO(patternQuery)
17:                     // add item in the result
18:                 end foreach
19:             else if queryS.IsVar && queryP.IsVar && queryO.IsValue then
20:                 foreach itemPattern in itemPatterns do
21:                     itemS ← Bindings.getS(patternQuery)
22:                     itemP ← patternQuery.getP
23:                     itemO ← Bindings.getO(patternQuery)
24:                     if itemObject == queryObject then
25:                         // add item in the result
26:                     end if
27:                 end foreach
28:             end if
29:         end foreach
30:     end foreach
31: end function
```

A comparison is made between the patterns of the query and those of the current item if at least one of the three elements (subject, predicate, object) of the pattern of the query is a constant. In this case, we compare the element(s) of the two patterns: if they are equal then we add the pattern of the current item in the result of the query by replacing the subject and the object with their values recovered in the bindings. Otherwise, we do not add the pattern of the item. This processing is carried out on all the grounds of the request before moving to the next pattern of the current item. After processing all the patterns of the current item, we move to the patterns of the next item.

Query with filter operator

Algorithm 2 filtering

1: **function** FILTER(*Items, query*)
2:     *filterType* ← *getFilter(query)*

3:     **if** *filterType is regexContains* **then**
4:         *regexContainsFilter(Items)*

5:     **else if** *filterType is numeric* **then**
6:         *numericFilter(ResultItem)*
7:     **end if**
8: **end function**
9: **function** REGEXCONTAINSFILTER(*Items*)
10:     *filterVar* ← *query.getFiterVar()*
11:     *filterValue* ← *query.getFilterValue()*

12:     **if** *varFilter is subject* **then**

13:         **while** *item in Items* **do**
14:             **if** *item.subject.contains(filterValue)* **then**
15:                 *// add item in the result filter*
16:             **end if**
17:         **end while**

18:     **else if** *filterVar is predicate* **then**

19:         **while** *item in Items* **do**
20:             **if** *item.predicate.contains(filterValue)* **then**
21:                 *// add item in the result filter*
22:             **end if**
23:         **end while**

24:     **else if** *filterVar is object* **then**

25:         **while** *item in Items* **do**
26:             **if** *item.object.contains(filterValue)* **then**
27:                 *// add item in the result filter*
28:             **end if**
29:         **end while**

30:     **end if**
31: **end function**
32: **function** NUMERICFILTER(*Items*)
33:     *filterValue* ← *query.getFilterValue()*
34:     *operator* ← *query.getFilterOperator()*

35:     **if** *operator is ε = ε* **then**

```
36:              while item in Items do
37:                  if item.object is filterValue then
38:                      // add item in result filter
39:                  end if
40:              end while

41:          else if operator is ε > ε then

42:              while item in Items do
43:                  if item.object > filterValue then
44:                      // add item in result filter
45:                  end if
46:              end while

47:          else if operator is ε < ε then

48:              while item in Items do
49:                  if item.object < valueFilter then
50:                      // add item in result filter
51:                  end if
52:              end while

53:      end if
54: end function
```

Filter operators allow to apply restrictions on numbers, strings, dates, etc. In our contribution, we have implemented the first two types of filters namely the filters on the numbers and those on the strings of characters. When a query contains a filter operator we check if it is an arithmetic expression filter or a string expression filter. If it is an arithmetic expression, we retrieve the two operands and the operators but if it is a string expression we only extract the two operands. The execution of the query is done in two steps: in the first one, we eliminate the filters and execute it as it was a simple query. We apply the filtering described in Algorithm 2. Depending on the type of filter (numeric or string expression), different data processing are performed in this job.

- **Arithmetic expression filter** The list of implemented arithmetic operations is: superiority, inferiority and equality. After executing the query on the n items, the result is filtered using the two operands and operators that were retrieved in the initialization phase. We go through the result returned by job 2 and for each triple we check and if it meets the conditions observed in the filter clause, we add it in the final result, otherwise, we eliminate it.
- **String expression filter** We implemented two string expression filtering i.e. the equality operator between two given strings and substring extraction (find a substring from a given string). It is the same principle as with the digital filters we go through the result returned by the job 2 and for each triple we check if it respects the conditions that are in the filter. If yes we add it in the final result, otherwise, we eliminate it.

Query with Aggregate Operator

Aggregate operators allow us to aggregate the result of a query. When the query to be executed contains aggregate operators, we retrieve the variables that are in the Group By clause and those on which the aggregate functions apply. The execution of a query with aggregate operators is done through two steps: in the first ones, we eliminate the aggregation and execute it as it was simple query. The second step concerns aggregation of the items and is described in Algorithm 3. Depending on the type (COUNT, SUM, AVG, MAX, MIN), different jobs are executed with the same principle.

Algorithm 3 aggregating

```
 1: function AGREGATOR( resultItem,query )
 2:     agregateFunction ← query.getAgregateFunction()
 3:     var ← query.getGroupByVar()

 4:     if agregateFunction is count then

 5:         if var is subject then
 6:             while result₁ in resultItem do
 7:                 while result₂ in resultItem do
 8:                     if result₁.subject equals result₂.subbject then
 9:                         count ← count + 1
10:                     end if
11:                 end while
12:                 // add result_1 subject and count number in the result
13:                 count ← 0
14:             end while
15:         else if var is predicate then
16:             while result₁ in resultItem do
17:                 while result₂ in resultItem do
18:                     if result₁.predicate equals result₂.predicate then
19:                         count ← count + 1
20:                     end if
21:                 end while
22:                 // add result_1 predicate and count number in the result
23:                 count ← 0
24:             end while
25:         else if var is object then
26:             while result₁ in resultItem do
27:                 while result₂ in resultItem do
28:                     if result₁.object equals result₂.object then
29:                         count ← count + 1
30:                     end if
31:                 end while
32:                 // add result_1 object and count number in the result
33:                 count ← 0
34:             end while
35:         end if
36:     end if
37: end function
```

We go through the result returned by job 2 of each window and we group the variables of the GROUP BY clause that we retrieved. Thus, for each group we apply the aggregate function. At the end, we display the result for each group.

With the contribution presented above, we were able to execute SPARQL queries on RDSZ based compressed RDF data. These queries can be of different forms. We implemented simple query execution with filter operators and aggregate operators. In the next section we will evaluate this contribution on different parameters in order to measure the performance.

5 Evaluation

In this section, we evaluate the performances of our contribution. We look at the processing time (data compression time and query execution time) and the consumed memory size. The results obtained are compared to those of the basic RDSZ algorithm. The evaluations are made on a computer with an AMD E2-1800 APU 1.70 GHz processor and 4 GB of RAM running with a Windows 10 system.

5.1 Data Provided

We use 9757 RDF graphs issued from observations of 20,000 weather stations. These dataset contains the aggregation of climatic data collected at various stations in the United States since 2002. These data are collected at the Meteorological Department of the University of Utah using the Kno.e.sis laboratory (The Ohio Center of Excellence in Knowledge enabled Computing).

5.2 In Terms of Execution Time

The execution time of a query depends on several parameters, namely the size of the batch, the size of the data, and so on. The size of the batch is the number of items that the system must receive before starting the data processing. We do the evaluation of the execution time by varying the size of the batch. To determine the performance of our contribution we are based on the execution time of a query with RDSZ. With RDSZ, we cannot directly querying compressed data using SPARQL query language.

The execution time of a request with RDSZ is thus the sum of the time needed for compression phase plus the one needed for decompression phase and the one needed for query execution. On the other hand, the execution time of a query with our contribution is the sum of the time needed for compression phase and the one needed for our execution model. Queries are executed on different batch sizes (100,

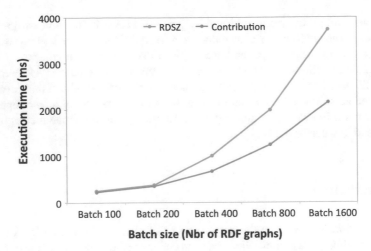

Fig. 3 Evaluation of execution time

200, 400, etc.) in order to see the evolution of time when the batch size increases. In the results (Fig. 3) we note that the larger the batch size increases the better the performance of the two curves decrease. This effect is due to the fact that when the size of the batch increases, the system must wait for a larger number of items to start its processing and consequently the processing time increases. The execution times with our contribution are smaller. Indeed, we execute the queries just after the compression step of RDSZ, which allows us to eliminate the decompression time. Therefore, we have smaller execution times.

5.3 In Terms of Memory Consumption

We evaluate the memory space used in our contribution based on RDSZ. The memory space used by RDSZ is calculated by taking the maximum of the memory space used between the compression and decompression phase and execution of the request for each item of a batch. So, to get the memory space used in a batch, we calculate the average. In the results (Fig. 4), we note that the larger the batch size increases the more memory space used. The memory space used by our contribution is smaller than the memory space used by RDSZ. Indeed, when executing the query with the basic RDSZ algorithm, the data is decompressed, thus occupying a part of the memory space allocated to the java virtual machine (jvm). By eliminating the decompression phase the space that should be used by this step is released.

We evaluated the performances of our contribution with two parameters: execution time and memory space. The tests we carried out were based on data with an overall size of 9757 RDF graphs. It should be noted that increasing the size of the test data leads to more gains in performance.

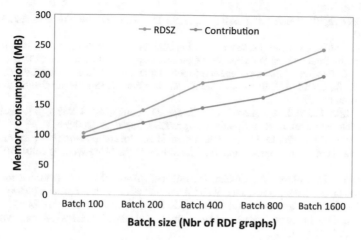

Fig. 4 Evaluation of memory consumption

6 Conclusion

In this paper, we proposed an extension of the RDSZ algorithm to allow continuous querying of RDF data streams in RDSZ format. This extension allowed us to execute SPARQL queries (simple, with filter and with aggregate) after the compression phase. The evaluation of this contribution has shown great gains on the executing time and consumed memory. As perspectives, we plan to take into account all SPARQL operators in order to take into account the temporal windows and to extend this work in order to be able to execute queries on the compressed data in the RDSZ format up to the binary level.

References

1. Abadi, D.J., Marcus, A., Madden, S.R., Hollenbach, K.: Scalable semantic web data management using vertical partitioning. In: Proceedings of the 33rd International Conference on Very Large Data Bases, pp. 411–422. VLDB Endowment (2007)
2. Álvarez-García, S., Brisaboa, N.R., Fernández, J.D., Martínez-Prieto, M.A.: Compressed k2-triples for full-in-memory rdf engines. arXiv preprint arXiv:1105.4004 (2011)
3. Anicic, D., Fodor, P., Rudolph, S., Stojanovic, N.: Ep-sparql: a unified language for event processing and stream reasoning. In: Proceedings of the 20th International Conference on World Wide Web, pp. 635–644. ACM (2011)
4. Barbieri, D., Braga, D., Ceri, S., Della Valle, E., Grossniklaus, M.: Stream reasoning: where we got so far. In: NeFoRS 2010: 4th International Workshop on New Forms of Reasoning for the Semantic Web: Scalable and Dynamic (2010)
5. Barbieri, D.F., Braga, D., Ceri, S., Della Valle, E., Grossniklaus, M.: C-sparql: Sparql for continuous querying. In: Proceedings of the 18th International Conference on World Wide Web, pp. 1061–1062. ACM (2009)

6. Berners-Lee, T., Hendler, J., Lassila, O., et al.: The semantic web. Sci. Am. **284**(5), 28–37 (2001)
7. Calbimonte, J.P., Corcho, O., Gray, A.J.: Enabling ontology-based access to streaming data sources. In: International Semantic Web Conference, pp. 96–111. Springer (2010)
8. Chiky, R.: Résumé de flux de données ditribués. Ph.D. thesis, Télécom ParisTech (2009)
9. Csernel, B., Clérot, F., Hébrail, G.: Classification de Flux de Donnes par chantillonnages sur Fentres Inclines
10. Della Valle, E., Ceri, S., Barbieri, D.F., Braga, D., Campi, A.: A first step towards stream reasoning. In: Future Internet Symposium, pp. 72–81. Springer (2008)
11. Fernández, J.D., Gutierrez, C., Martínez-Prieto, M.A.: Rdf compression: basic approaches. In: Proceedings of the 19th International Conference on World Wide Web, pp. 1091–1092. ACM (2010)
12. Fernández, J.D., Llaves, A., Corcho, O.: Efficient rdf interchange (eri) format for rdf data streams. In: International Semantic Web Conference, pp. 244–259. Springer (2014)
13. Fernández, J.D., Martínez-Prieto, M.A., Gutiérrez, C., Polleres, A., Arias, M.: Binary RDF representation for publication and exchange (hdt). Web Semant. Sci. Serv. Agents World Wide Web **19**, 22–41 (2013)
14. Fernández, N., Arias, J., Sánchez, L., Fuentes-Lorenzo, D., Corcho, Ó.: RDSZ: an approach for lossless RDF stream compression. In: European Semantic Web Conference, pp. 52–67. Springer (2014)
15. Joshi, A.K., Hitzler, P., Dong, G.: Logical linked data compression. In: Extended Semantic Web Conference, pp. 170–184. Springer (2013)
16. Komazec, S., Cerri, D., Fensel, D.: Sparkwave: continuous schema-enhanced pattern matching over RDF data streams. In: Proceedings of the 6th ACM International Conference on Distributed Event-Based Systems, pp. 58–68. ACM (2012)
17. Le-Phuoc, D., Dao-Tran, M., Parreira, J.X., Hauswirth, M.: A native and adaptive approach for unified processing of linked streams and linked data. In: International Semantic Web Conference, pp. 370–388. Springer (2011)
18. Urbani, J., Maassen, J., Drost, N., Seinstra, F., Bal, H.: Scalable RDF data compression with mapreduce. Concurr. Comput. Pract. Exp. **25**(1), 24–39 (2013)

Energy Efficiency Cluster Head Election using Fuzzy Logic Method for Wireless Sensor Networks

Wided Abidi and Tahar Ezzedine

Abstract The main challenge in wireless sensors networks (WSN) is to conserve the energy consumption and prolong the lifetime of network. Since sensor nodes are deployed in hostile area and it is difficult to recharge their batteries or change it, we must maintain the lifetime of these nodes as longer as possible. Electing the appropriate Cluster Head (CH) becomes very important. Many clustering algorithms have been developed for selecting the best CHs. In this paper, we introduce a new clustering algorithm which elects CHs using fuzzy logic method and based on a set of parameters which increases the lifetime of WSN. In fact, we adopt three principle criteria: the remaining energy of node, the number of neighbors within cluster range and the distance between node and CH for electing best suitable nodes as CH. Simulation results shows that our proposed algorithm beats the other algorithms in regards of prolonging the lifetime of network and saving residual energy.

Keywords Wireless sensors networks · Fuzzy logic · Clustering · Cluster head election · Network lifetime

1 Introduction

Wireless Sensor Network (WSN) consists of large number of tiny devices called sensor nodes [1]. These Nodes are deployed randomly in a geographical area. Their roles are to sense, collect, aggregate and send data between each other or to a Base Station (BS) located outside of the sensor area. This communication costs important energy consumption. On the other hand, sensor nodes use batteries as power source

W. Abidi (✉) · T. Ezzedine
Engineering School of Tunis, Communications Systems Laboratory,
University of Tunis El Manar, Tunis, Tunisia
e-mail: abidiwided@gmail.com

T. Ezzedine
e-mail: taharezz@gmail.com

© Springer International Publishing AG 2018
R. Lee (ed.), *Software Engineering Research, Management and Applications*,
Studies in Computational Intelligence 722, DOI 10.1007/978-3-319-61388-8_10

that are limited resources. In addition, this power source is usually not replaceable or rechargeable. Hence, the need to extend the lifetime of nodes and minimize the energy consumption is necessary.

Due to the energy constraints of the large number of deployed sensors, gathering nodes into groups called Clusters becomes very challenging. In each cluster, there is only one node which is allowed to communicate with the BS called Cluster Head (CH). Its main role is to collect the data sent by each node into cluster then transmits aggregated information to BS. Many clustering algorithms have been developed for WSN [2–4] which elects CHs through probabilistic approach or deterministic approach particularly weight based approach. Probabilistic approach elects CH without taking into account several lifetime factors like remaining energy of the node, the number of neighbor nodes and the distance between CH and BS or node and CH. The weight based approach considers the factors cited previously. But its main shortcoming is frequently same nodes are elected as CH which causes the loss of energy and subsequently the rapid death of the CH.

Since deficiencies of these two approaches, researchers have used fuzzy Logic method to optimize CH selection. In reality, fuzzy logic method has been developed to model the human decision making behavior. However, researchers used it to divide the WSN to clusters with best set of CHs.

The fuzzy logic model is divided to four components as illustrated in Fig. 1: a fuzzifier, fuzzy inference engine, fuzzy rule base and a defuzzifier. The role of the fuzzifier is to convert crisp value to fuzzy input variables. Fuzzy rule base stores IF-THEN rules. Based on these rules, fuzzy inference engine maps the set input linguistic variables to the output linguistic variables. At the last, the defuzzifier converts fuzzy output to crip output using the appropriate defuzzification method.

In this paper, we propose a new approach for electing CH using the fuzzy logic method. Our algorithm is based on remaining energy of the node, the distance between the node and the Base Station (BS) and the number of neighbor nodes within the range to elect CH. Thus, elected CH must have at the same time a high residual energy, maximum number of neighbor and finally a low distance to sink. By considering these factors, we can save energy consumption and prolong the lifetime of the network and good results will be shown by simulations later in the paper.

Fig. 1 Block diagram of fuzzy inference system

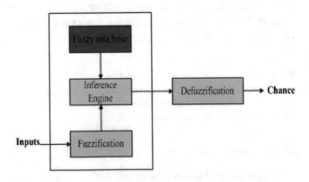

The rest of our paper is organized as follows. Clustering algorithms for WSN are presented in related work in Sect. 2. Section 3 details the proposed algorithm to select CH. Simulation results are shown and discussed in Sect. 4. We conclude in Sect. 5.

2 Related Works

In the literature, there are several proposed clustering algorithms for WSN. In this section, we will give an overview on the famous and recent one.

Low Energy Adaptive Clustering Hierarchy (LEACH) is a distributed clustering protocol. It has been introduced by [5] to reduce power consumption and prolong the lifetime of the WSN. LEACH divides network to clusters and only one node (CH) in each cluster is the leader and it changes each round. CH communicates directly with the BS to send data and uses data aggregation technique what reduce energy consumption. LEACH consists of two main phase: cluster setup phase and steady-state phase. In cluster setup phase, each node in the network decides if it will be a Cluster Head (CH) or not for present round. This decision depends on the desired percentage of CHs in the network and the number of times the node is served as CH so far. In fact, each node chooses a random number between 0 and 1. If this number is less than a threshold T, the node becomes a CH. Then, each CH broadcasts advertising messages to the remaining nodes inviting it to choose which of the CHs they will join and finally, clusters are created for the current round. Based on the number of nodes in the cluster, the CH creates a time division multiple access (TDMA) schedule and informs other sensor nodes when it can transmit. Finally, in steady-state phase, transmission data starts. Sensor nodes send their data in their own time slot and their radio can be turned off. CH must keep their radio on to receive all data from nodes.

Cluster head election using fuzzy logic (CHEF) [6] is a distributed clustering algorithm that uses fuzzy logic approach for WSN. When starting the round, CHEF elects candidate CHs using the same probability approach as LEACH protocol. Then it calculates the chance of these candidate CHs using the fuzzy logic method and based on residual energy and local distance of the nodes as input. The local distance is the sum of distance from all one-hop neighbors to a node. Fuzzy if-then rules are used to evaluate the fuzzified input values. The output variable chance arbitrates which candidate CH becomes a CH. The elected CH has higher chance value than its candidate CHs neighbors. The input fuzzy variable local distance is not a suitable variable for all network sizes and because of this, CHEF suffers a lot in network size apart from 200 m × 200 m.

Energy aware unequal clustering using fuzzy approach (EAUCF) [7, 8] is also a distributed clustering algorithm for WSN. When starting the round, tentative CHs are selected by using random number generation such as CHEF algorithm. Fuzzy system in EAUCF is based on two inputs: residual energy and distance to BS. The output is the competition radius which calculated to each tentative CH node.

Each tentative CH will broadcast its residual energy and check the existence of any other tentative CH node within its competition radius. If two such tentative CHs are present within the competition radius of either one node, the nodes having lesser residual energy will quit from CH competition. In CH election, an important parameter like node degree is not considered which may lead to election of CH with fewer and distant neighbors. These results in higher intra cluster communication cost and reduces the lifetime of the WSN.

3   Proposed Clustering Algorithm

3.1   System Assumptions

- Homogenous network is assumed where all nodes are having equal capabilities in terms of processing power, sensing area, and so forth.
- Sensor nodes are deployed randomly.
- Once deployed, nodes are static.
- All sensor nodes have the same initial energy.
- The base station is located in the outside of the WSNs.

3.2   Fuzzy System Model

Given that CH node has many activities to accomplish such as collecting information from other nodes, aggregating it and sending it to the BS, CH needs more energy level than member node. In other hand, the number of alive neighbors of a node within the radius R called neighbor nodes Alive is a factor which determines how a node is located. The energy consumption for transmitting data increases when the distance between transmitter and receiver nodes increases. From an energy conservation perspective, the distance between CH and BS should be minimized. Based on these factors, we proposed a clustering algorithm that considers the remaining energy (E_{rem}) of the node, the number of neighbor nodes (NA) and the distance between node and BS (d_{toBS}) to select the CH. This proposed algorithm used the fuzzy logic method.

Our fuzzy system is composed of three fuzzy inputs and one fuzzy output. For the first fuzzy input E_{rem}, the fuzzy linguistic variables used are Low, Medium, and High as depicted in Fig. 2. For the second fuzzy input NA, there are three fuzzy linguistic variables which are Several, Medium and Few as shown in Fig. 3. Finally, for the fuzzy input d_{toBS}, the fuzzy linguistic variables are Close, Average, and Far as illustrated in Fig. 4.

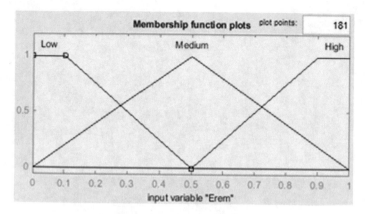

Fig. 2 Fuzzy set for the input variable remaining energy

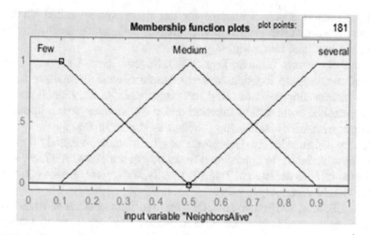

Fig. 3 Fuzzy set for the input variable neighbors alive

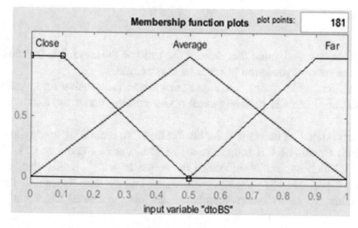

Fig. 4 Fuzzy set for the input variable distance to BS

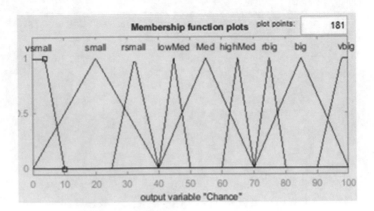

Fig. 5 Fuzzy set for the output variable chance

There is only one output variable for our fuzzy system called Chance. This output has nine linguistic variables: very small, small, rather small, low medium, medium, big medium, rather big, big, and very big.

Note that for inputs variable: Low, High, Several, Few, Close and Far follow trapezoidal membership function, whereas Medium, Medium and Average follow triangular membership function. And for output variable, very small and very big follow trapezoidal membership function and the remaining seven linguistic variables follow triangular membership function as shown in Fig. 5.

The fuzzy if-then rules are developed based on Mamdani method [9] and used to map the input variables to appropriate fuzzy output variables. We have at total 27 fuzzy if-then rules as depicted in Table 1. Finally, the Center of Area (CoA) method is used to obtain crisp output values.

Such as LEACH, CHEF and EAUCF, our algorithm operates in round. In every round, each sensor node chooses a random number between 0 and 1. If this number is less than a threshold P_{opt} calculates with Eq. (1), the node becomes a Candidate CH.

$$P_{opt} = \alpha * P \tag{1}$$

Where α is a constant value that defines the ratio of the candidate for cluster head and P is the ratio of preferred number of cluster heads.

The Candidate CH nodes calculate their chance value using fuzzy method and broadcast Candidate CH message to all nodes coming under their communication radius.

The Candidate CH message contains the node id, remaining energy and chance value. Each Candidate CH broadcast to the total list of Candidate CH. The Candidate CH which has the high chance is elected as a CH. If this elected CH has received a Candidate CH message with higher chance than its own. It becomes a node member. To form clusters, member nodes join the nearest CH. Finally, elected

Table 1 Fuzzy if-then rule

Remaining energy	Neighbors alive	Distance to BS	Chance
Low	Several	Close	Rather small
Low	Medium	Close	Small
Low	Few	Close	Very small
Low	Several	Average	Rather small
Low	Medium	Average	Small
Low	Few	Average	Very small
Low	Several	Far	Rather small
Low	Medium	Far	Small
Low	Few	Far	Very small
Medium	Several	Close	High med
Medium	Medium	Close	Med
Medium	Few	Close	Low med
Medium	Several	Average	High med
Medium	Medium	Average	Med
Medium	Few	Average	Low med
Medium	Several	Far	High med
Medium	Medium	Far	Med
Medium	Few	Far	Low med
High	Several	Close	Very big
High	Medium	Close	Big
High	Few	Close	rather big
High	Several	Average	very big
High	Medium	Average	big
High	Few	Average	Rather big
High	Several	Far	Very big
High	Medium	Far	Big
High	Few	Far	Rather big

CHs generate TDMA schedule for their members and broadcast it. Then, each member nodes send their data to its CH during their allocated time slots. Otherwise, they go to sleep state to save energy.

3.3 Energy Model

In our research, we have used the same energy model as the traditional LEACH [10], as shown in "Fig. 6." Note that, E_{elec} is the energy consumption per bit for running transmitter or receiver circuitry, k is the number of bits, ε_{fs} and ε_{mp} are proportional constant of the energy consumption for the transmit amplifier in free space channel model ($\varepsilon_{fs} \cdot k \cdot d^2$ power loss) and multipath fading channel model

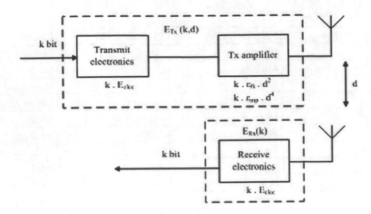

Fig. 6 The radio energy consumption model

($\varepsilon_{mp} \cdot k \cdot d^4$ power loss), respectively and d is the distance between transmitter and receiver.

Thus we can deduce the energy consumed to transmit k bits along a distance d through a free space channel model is:

$$E_{Tx}(k, d) = E_{elec} * k + \varepsilon_{fs} * k * d^2 \tag{2}$$

Or multipath fading channel is:

$$E_{Tx}(k, d) = E_{elec} * k + \varepsilon_{mp} * k * d^4 \tag{3}$$

And the energy to receive these bits is:

$$E_{Rx}(k) = E_{elec} * k \tag{4}$$

4 Simulations and Numerical Results

In this section, simulations are performed via Matlab software in the same conditions. We have compared between our proposed approach and LEACH protocol using parameters listed in Table 2. We consider a WSN with randomly distributed sensor nodes in 100 × 100 network field. BS is located at the coordinate (50, 175).

Simulation results are analyzed with considering the First Node Dies (FND), Half Nodes Die (HND) and the remaining energy of the network per round.

Figures 7 and 8 shows network lifetime. Our proposed algorithm shows increased FND compared to LEACH, CHEF and EAUCF algorithms. As depicted in Table 3, considering FND, our proposed algorithm increases network lifetime compared to LEACH by 36%, CHEF by 14% and EAUCF by 12%. On considering

Table 2 Parameters system

Simulation area	$100 \times 100 \text{ m}^2$
Number of round	1000
Number of nodes	200
Desired percentage of CH	0.1
Initial energy of node	1 J
Transmission/Reception energy per bit E_{elec}	50 nJ/bit
Transmitter amplifier energy dissipation free space ε_{fs}	10 pJ/bit/m^2
Transmitter amplifier energy dissipation multipath ε_{mp}	0.0013 pJ/bit/m^4
Base station location	Located at 50×175
P_{opt}	0.21

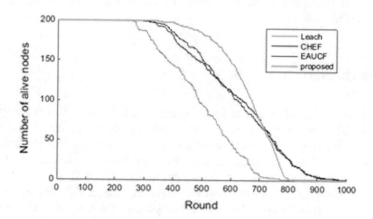

Fig. 7 The number of alive nodes per round

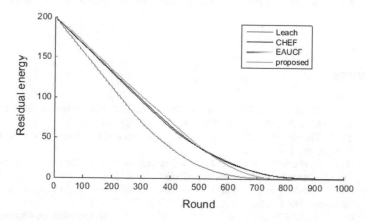

Fig. 8 Total residual energy per round

Table 3 FND and HND for 200 nodes

Algorithms	FND	HND
LEACH	264	488
CHEF	314	641
EAUCF	320	624
Proposed algorithm	358	672

HND, our proposed algorithm is better than LEACH by 38%, CHEF by 4% and EAUCF by 8%.

Figure 8 shows the remaining energy for the network per round. For LEACH protocol, remaining energy is consumed more quickly than other algorithms. The remaining energy levels of CHEF and EAUCF are close to each other. However, for our proposed algorithm, remaining energy increases slightly slower than LEACH, CHEF and EAUCF.

5 Conclusions

In WSN, the main purpose is to increase network lifetime and save energy consumption. This paper proposed a new clustering algorithm using fuzzy logic method and based on a set of important parameters which influence the lifetime of the network. In fact, our proposed algorithm combines probabilistic and metric based CH election techniques with suitable criteria for CH election in WSN. Simulation results are presented comparing with basic LEACH, CHEF and EAUCF algorithms and it is observed that our developed approach is more effective than LEACH in reducing energy consumption and extending lifetime network.

In future work, more parameters like density and number of hops between CH and BS may also be utilized to further improve the performance of our fuzzy algorithm.

References

1. Anastasi, G., Conti, M., Di Francesco, M., Passarella, A.: Energy conservation in wireless sensor networks: a survey. Ad Hoc Netw. J. **7**(3), 537–568 (2008)
2. Afsar, M.M., Tayarani-N, M.H.: Clustering in sensor networks: a literature survey. J. Netw. Comput. Appl. **46**, 198–226 (2014)
3. Singh, S.K., Singh, M.P., Singh, D.K.: Routing protocols in wireless sensor networks—a survey. Int. J. Comput. Sci. Eng. Surv. (IJCSES) **1**(2) (2010)
4. Katiyar, N.V., Chand, Soni, S.: A survey on clustering algorithms for heterogeneous wireless sensor networks. Int. J. Adv. Netw. Appl. **2**(4), 745–754 (2011)
5. Heinzelman, W.B., Chandrakasan, A., Balakrishnan, H.: Energy-efficient communication protocol for wireless microsensor networks. In: Proceedings of the 33rd Annual Hawaii

International Conference on System Sciences (HICSS'00), vol. 8, pp. 1–10. Maui, Hawaii, USA (2000)

6. Kim, J.M., Park, S.H., Han, Y. J., Chung, T.M.: CHEF: cluster head election mechanism using fuzzy logic in wireless sensor networks. In: Proceedings of the 10th International Conference on Advanced Communication Technology, pp. 654–659. Gangwon-Do, South Korea (2008)

7. Bagci, H., Yazici, A.: An energy aware fuzzy approach to unequal clustering in wireless sensor networks. Appl. Soft Comput. J. **13**(4), 1741–1749 (2013)

8. Bagci, H., Yazici, A.: An energy aware fuzzy unequal clustering algorithm for wireless sensor networks. In: Proceedings of the 6th IEEE World Congress on Computational Intelligence (WCCI'10). IEEE (2010)

9. Mamdani, E.H.: Application of fuzzy logic to approximate reasoning using linguistic synthesis. IEEE Trans. Comput. **26**(12), 1182–1191 (1977)

10. Heinzelman, W.B.: Application specific protocol architectures for wireless networks. Ph.D. Thesis, Department of Electrical Engineering and Computer Science, Massachusetts Institute of Technology, Cambridge, MA (2000)

Enabling GSD Task Allocation via Cloud-Based Software Processes

Sami Alajrami, Barbara Gallina and Alexander Romanovsky

Abstract Allocating tasks to distributed sites in Global Software Development (GSD) projects is often done unsystematically and based on the personal experience of project managers. Wrong allocation decisions increase the project's risks as tasks have dependencies that are inherited by the distributed sites. Decision support can help make the task allocation a more informed and systematic process. The challenges in allocating tasks to distributed sites exist because of three distance dimensions between sites (geographical, temporal and cultural). An informed task allocation decision needs to consider these distances. Therefore, in this paper, we propose to integrate and semi-automate the calculation of an existing Global Distance Metric (GDM) into an architecture that supports executing cloud-based software processes. We analyze the potential of integrating the GDM into this architecture and identify the needed extensions to the architecture.

Keywords Global software development · Distributed tasks allocation decision support · Cloud-based software processes · Global distance

1 Introduction

Global Software Development (GSD) [11] has moved software firms from monolithic development (one team at one location) to multiple geographically-distributed teams collaborating on a development project. GSD benefits are established in literature [6, 8, 11] and include: (a) utilizing cheaper labour in different countries hence implying cost reduction, (b) having multiple teams working in different time zones

S. Alajrami (✉) · A. Romanovsky
Newcastle University, Newcastle upon Tyne, UK
e-mail: s.h.alajrami@ncl.ac.uk

A. Romanovsky
e-mail: alexander.romanovsky@ncl.ac.uk

B. Gallina
Mälaradalen Univeristy, Västerås, Sweden
e-mail: barbara.gallina@mdh.se

© Springer International Publishing AG 2018
R. Lee (ed.), *Software Engineering Research, Management and Applications*,
Studies in Computational Intelligence 722, DOI 10.1007/978-3-319-61388-8_11

which leads to shorter development cycles, and (c) being in closer proximity to customers and emerging markets.

Despite the benefits, teams collaborating in GSD projects face geographical, temporal and cultural distances which make managing such projects a challenging task. Naturally, dependencies exist between the distributed tasks. These task dependencies (during process enactment) make it essential to ensure no deadlocks happen between distributed sites.

The distances between distributed sites introduce management challenges that can increase the risks for GSD projects. Such management challenges are inherent in GSD projects and are linked to issues of communication, control and supervision, coordination, creating social bonds, and building trust [7]. Among the main GSD challenges, allocating the right resources/tasks to each site is of critical importance.

The complexity of the dependencies in GSD projects is reflected on the task allocation decisions [12]. Task allocation can either decrease or increase the risks associated with GSD projects (such as: decreased productivity and lack of trust between sites) [13]. Despite the importance of task allocation decisions, in practice, the decision making process is not very systematic and often is based on the personal experience of the managers [14]. For example, allocating activities to sites with low differences (nearshoring [7]) seems to reduce GSD risks, while having large cultural differences between sites affects the trust between them. Therefore, a systematic decision support is needed to support allocating GSD activities.

The larger the distance between distributed sites, the larger the difference. Nearshoring [7] (allocating tasks to sites with low differences) reduces the risks associated with GSD projects management [13]. Carmel and Abbott argue that the rise of nearshoring proves that distance still matters [7]. Therefore, in this paper, we explore how we can make informed decisions about task allocation in GSD projects based on the distances between the distributed sites. In order to base the decision making on the distance factor, this factor needs to be quantified. For that purpose, we use the *Global Distance Metric* [17] which assesses and quantifies the distance between collaborating sites.

In a previous work, we proposed a reference architecture for supporting Software Development as a Service (SDaaS) in the cloud [5]. The potential for using the cloud to facilitate GSD projects has been discussed in [10]. The SDaaS architecture goes one step further and uses a model-based approach to execute software processes (which can be distributed processes). The SDaaS architecture facilitates by default: global project awareness, enhancing communication and understanding amongst distributed teams and supporting global monitoring and synchronization of tasks. In addition, executable process models (when supported with the appropriate execution environment) can help addressing technical GSD challenges such as: incompatible data formats and tools [2]. Therefore, in this paper, we propose to extend the SDaaS architecture to support semi-automatic calculation of the *Global Distance Metric* in order to provide task allocation decision support for project managers.

The rest of the paper is structured as follows: Section 2 provides brief background on the SDaaS architecture and Global Distance Metric (GDM). Section 3 describes our proposed extension of the SDaaS architecture to provide GSD task allocation decision support. Section 4 explains the paper proposal using an example process. Section 5 reviews some existing works that target task allocation support in GSD projects. Finally, Sect. 6 concludes the paper and discusses the current limitations.

2 Background and Motivation

In this section, we briefly cover essential background information on our architecture for executing cloud-based software processes and on GSD distance metric and task allocation.

2.1 The SDaaS Architecture

We proposed a reference architecture for supporting executing software process models in the cloud [5]. As shown in Fig. 1, the architecture consists of two main services: the design time service and the run-time service. The design time service deals with modelling and manipulation of software processes while the run-time service deals with scheduling, executing and monitoring software processes execution in the cloud. The execution takes place in a set of distributed workflow engines (with dif-

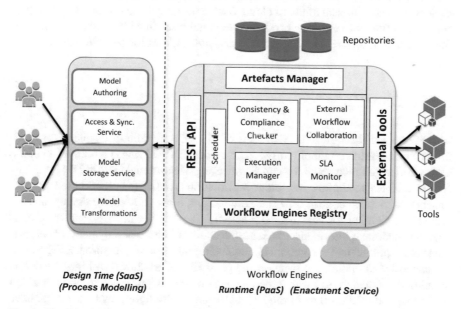

Fig. 1 The SDaaS reference architecture. Taken from [5]

ferent computational and privacy specification). The workflow registry component tracks and manages the active workflow engines. During the execution, process models consume and produce software artefacts (code, docs, models, tests etc.). These artefacts are maintained along with meta-data describing them by the artefact manager component. The tools needed to support each process activity can be integrated within the environment or can be interfaced as a service. Activities can be: (a) automated (triggering tools to perform certain tasks e.g., testing), (b) interactive (receiving input from users e.g., for editing artefacts), or (c) decision points (deciding—automatically or interactively—on which branch of the process to follow).

SDaaS facilitates distributed development. It uses a unified SaaS user interface which enables teams across distributed sites to access a shared development environment. This means that teams will be collaborating within the same virtual environment which is highly accessible and available via the cloud. The cloud model is based on provisioning of services and the SDaaS architecture provisions development environments and tool-chains as services. Hashmi et al. [10] argue that GSD challenges can be overcome via the use of services (Service Oriented Architecture - SOA). Their argument is that SOA increases the interoperability and technology and business alignment between sites [10]. Since the SDaaS architecture adopts a SOA, we argue that it can overcome GSD challenges.

In addition, the SDaaS architecture adopts a model-driven approach and supports modelling of dynamic processes like the ones that would be found in GSD projects. The use of models allows for raising the levels of abstraction and improves communication and understanding between distributed sites. The artefact manager of the SDaaS architecture allows for tracing and maintaining shared artefacts. Finally, SDaaS leverages the scalability of cloud to allocate computing resources and tools as services on the fly to meet the needs of individual tasks in a GSD project. However, the SDaaS architecture does not provide decision support for task allocation.

2.2 EXE-SPEM

The SDaaS architecture uses EXE-SPEM [3] as the modelling language for modelling cloud-based executable software processes. EXE-SPEM is an extension of the OMG Software Process and System Engineering Meta-model (SPEM2.0 [1]). EXE-SPEM enables modelling important information needed for cloud-based process enactment such as: control flow (i.e., order, conditions and loops), the responsible team/team member for enacting each activity (task) in the process, and the cloud-specific enactment information such as: the choice of cloud deployment model (private versus public) and the amount of computational resources required. EXE-SPEM is created by extending the meta-model of SPEM2.0 as shown in Fig. 2 (which is simplified for readability) where meta-classes with dark grey background are added to the original SPEM2.0 meta-model and the ones with light grey background have new attributes.

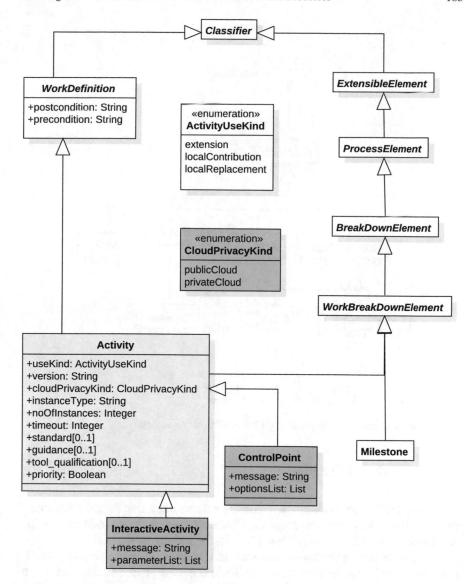

Fig. 2   The meta-model of EXE-SPEM

Using model-to-text transformational rules, EXE-SPEM process models are mapped into XML-based textual representations which are compliant with the schema shown in Fig. 3.

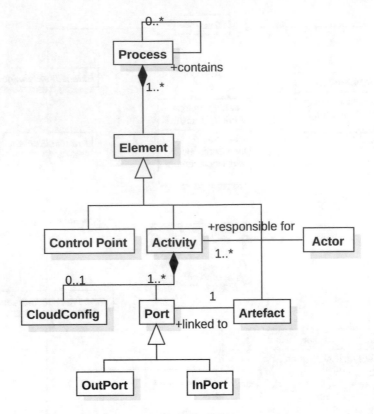

Fig. 3 The XML schema for representing EXE-SPEM process models

2.3 GSD Task Allocation

Allocating GSD tasks to distributed sites has a direct impact on the risks associated with distributed development projects. Allocation is often done based on multiple criteria (labor cost rates, availability of workforce and expertise) [13].

Lamersdorf et al. have reviewed several tactics followed in practice to avoid the risks associated with distance between distributed sites [14]. The first tactic is to minimize the collaboration needed (separation of concerns between sites) which reduces the GSD communication problems. Another tactic is to minimize the differences (e.g., cultural, temporal) between sites. Grinter et al. [9] proposed the use of strategies from organizational theory to task allocation in GSD projects.

The optimal task allocation decision needs to be based on understanding of the capabilities, differences and distances among the distributed tasks. Distance between sites is the main source of risk in GSD projects and it takes different dimensions (geographical, temporal and cultural). Thus, quantifying these dimensions of distance helps to make an effective and informed task allocation decision by project managers.

2.4   Global Distance Metric

Noll and Beecham [17] have developed the global distance metric (GDM) to measure global distance between distributed sites collaborating on GSD projects. The metric combines and quantifies the three dimensions of GSD distance: geographic, temporal, and cultural between two sites. The metric is then calculated as follows:

$$D_{global} = \sqrt{D^2_{geographic} + D^2_{temporal} + D^2_{cultural}} \tag{1}$$

where D_c is the value of the distance dimension and $c \in \{geographic, temporal, cultural\}$. Each of the dimensions in Eq. 1 is calculated as the sum of the impact values for different distance factors. A list of these factors and there impact values is provided in Table 1. Each team (site) computes the global distance metric from other collaborating sites. This provides a quantified representation of the perceived distances between the distributed sites towards each other.

Table 1 is taken from [17] and shows the factors contributing to each distance dimension along with their impact values. These impact values have been identified by surveying practitioners. As we can see in the table, factors affecting both the geographical and temporal distances are straightforward to assess (based on the locations and timezones of distributed sites). However, the cultural distance depends more on the perception and trust between teams. For example, as noted by Noll and Beecham [17], having a team member from the same nationality (of a certain site) in another site may lead to increase the perceived trust and reduce the perceived language barriers.

Table 1   Factors contributing to distances [17]

No.	Factors affecting geographic distance	Impact value
1	Different building on same campus	1
2	Different towns in same region (two hour drive)	2
3	Less than three hour flight (Frankfurt to Helsinki)	3
4	Transcontinental flight (New York to San Francisco)	4
5	Intercontinental flight (London to Shanghai)	4
No.	Factors affecting temporal distance	Impact value
1	Transcontinental (five hour overlap)	0
2	Intercontinental (three or four hour overlap)	3
3	Global (one or two hour overlap)	4
4	No overlap	4
No.	Factors affecting cultural distance	Impact value
1	Uneven language skills	3
2	East/West divide in culture	3
3	Different national culture	2
4	Different organizational culture	3

3   SDaaS-Based Task Allocation

In this section, we build on existing GSD support in the SDaaS architecture by facilitating decision making about allocating tasks across distributed sites. Since knowing the distance (in all its dimensions) between distributed sites is crucial for making the right allocation decision, we propose to integrate the measurement of the *Global Distance Metric (GDM)* [17] (see Sect. 2.4) within the SDaaS architecture.

The SDaaS architecture can automate the measurement of the geographical and temporal distances of the GDM based on knowing the collaborating sites and their locations. In addition, it can calculate the cultural distance perceived by each site towards each other site by relying on input from team members. These calculated values can then be used to calculate the overall GDM between each two sites using Eq. 1.

3.1   The SDaaS Architecture Extension

In order to support the GDM calculation, the SDaaS architecture needs to be extended. Task allocation is needed during the process design-time phase. The following extensions are needed in the SDaaS architecture:

1. Extending the process models

   The teams which might be involved in executing the process and their respective sites need to be integrated in the process models (which are created using EXE-SPEM [3]). We extend the EXE-SPEM meta-model which defines EXE-SPEM process models elements. As shown in Fig. 4, the extended meta-model of EXE-SPEM integrates the *Site* and *Team* meta-classes (in dark grey). The *Activity* meta-class has a new attribute stating the site that the activity has been allocated to. Finally, the *CulturalDistanceKind* enumeration is added to represent different cultural distance factors as shown in Table 1.

   In addition to extending the meta-model of EXE-SPEM, we extend the schema for defining the XML representation of EXE-SPEM process models as shown in Fig. 5 where the *Site* and *Team* have been added.

2. Adding a GDM calculation module

   The design-time part of the SDaaS architecture (see Fig. 1) needs to be extended by adding a module for calculating the GDM (following Eq. 1). The geographical and temporal distance factors can be automatically calculated by this module using the team and site information from the process model. The cultural distance, however, is a subjective factor. Therefore, this module should interact with the team members to calculate their perceived cultural distance factors towards other teams at different sites. This can be done using the factors from Table 1.

3. Visualizing the GDM between distributed sites

   Once the GDM between each pair of distributed sites is calculated, the project manager needs to view the overall perceived distances between distributed sites

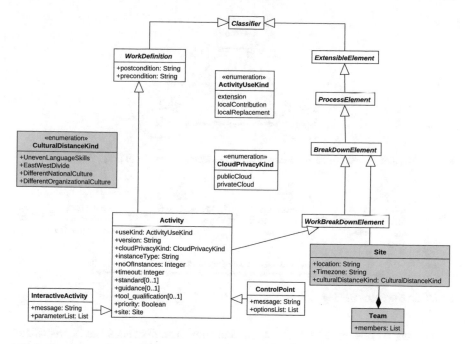

Fig. 4 The extended meta-model of EXE-SPEM

Fig. 5 The software process
model XML schema

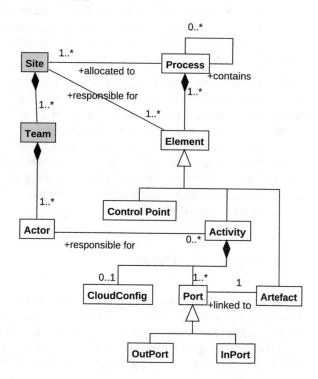

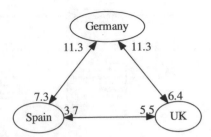

Fig. 6 The global distance between three distributed teams. Taken from [17]

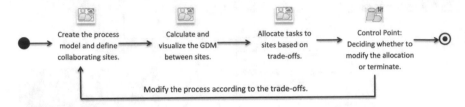

Fig. 7 The decision making process

in order to make the best allocation decisions. The distances can be visualized following the example in Fig. 6 which is taken from Noll and Beecham [17] and shows the distances between three distributed teams (Germany, Spain and UK). The numbers represent the perceived distance from one site towards another. The larger the number, the larger the distance and consequently, the larger the differences and risks.

The decision making process is depicted in Fig. 7. It starts with the project manager or process author creating the process model and specifying the teams that might be involved in this process. Then, the GDM between these sites is calculated and visualized. Finally, the project manager makes a decision to allocate specific tasks to specific teams based on a trade-off between multiple factors (e.g., labour cost, availability, expertise and GDM). Based on the trade-offs, the project manager may decide to make modifications to the process in order to reduce the risks associated with involvement of distributed teams. For example, to reduce dependencies between two teams with high GDM value.

4 Demonstrating Example

To demonstrate the proposed approach in this paper, we use a process model we developed in a previous work [4]. The process is a safety process for generating product and process safety arguments to be used in building safety cases for safety critical systems. Figure 8 shows the original process (before introducing the extension for task allocation support) modelled in EXE-SPEM. The process consists of

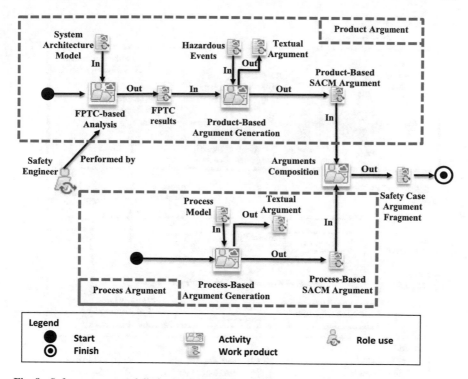

Fig. 8 Safety process modelled using EXE-SPEM. Adapted from [4]

activities which consume and produce *work products* (artefacts) and are performed by *role use* (actors).

Figure 9 shows the same process modelled with the extended EXE-SPEM. As the figure shows, the model now describe the collaborating sites (one in the UK and another in India). By analyzing calculating the GDM between these two sites from the process model, the distances can be reported and visualized to project managers who can then make an informed decision to allocate certain activities to certain sites. For example, as shown in Fig. 9, the decision could be to allocate the *Product-based Argument Generation* activity to the UK site and the *Process-based Argument Generation* to the Indian site. After allocating the activities to sites, the process model can be executed in the SDaaS architecture.

5 Related Work

Several approaches for task allocation in GSD projects have been studied in literature. Some studies have reviewed these approaches (e.g. [12, 14]). Imtiaz and Ikram [12] have identified several factors that impact task allocation in GSD projects such as: labour cost, expertise, task-site dependency, temporal and cultural dif-

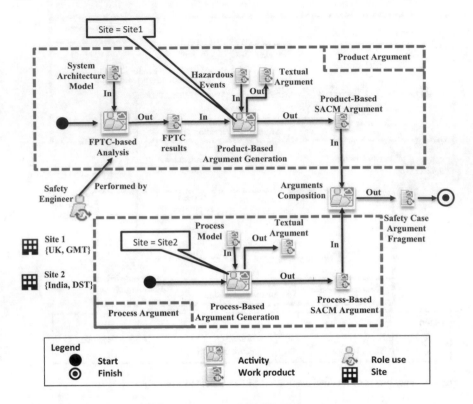

Fig. 9 Safety process modelled using the extended EXE-SPEM

ferences, etc. Task allocation approaches often target one or few of these factors and a trade-off between them need to be performed based on the situation and the project [12].

Task allocation for GSD projects can be categorized into two groups [16]: (a) *optimization approaches* (aiming to decide on the best task allocation with respect to a specific goal) and (b) *predictive approaches* (aiming to evaluate different task allocations individually).

Mockus and Weiss [15] propose an optimization algorithm which aims to minimize the communication needed between sites and thus reducing the communication overhead. However, this approach only addresses a single criterion (i.e., communication overhead). Another approach developed by Setamanit et al. [18] uses a simulation model to compare different task allocation strategies with respect to productivity and development time. This approach, however, does not provide task allocation decision support for individual projects and instead compare the strategies generally. Lamersdorf and Münch [13] study the risk identification and effort estimation perspectives in GSD task allocation and conclude that although some approaches can be used to support certain aspects of task allocation, there is no comprehensive approach for systematic task allocation covering all the needed aspects.

6 Conclusion and Future Work

In this paper, we extend the SDaaS architecture [5] to provide task allocation decision support for GSD projects. SDaaS facilitates conducting GSD projects in the cloud and automate the computational ad tool resources allocation on demand. The extension uses the Global Distance Metric (GDM) [17] to quantify the three dimensions of GSD distance (geographical, temporal and cultural). This extension allows projects managers to make task allocation decisions baring in mind the distances (differences) between the collaborating distributed tasks and the risks associated with it.

In practice, the decision on task allocation is made based on multiple factors (e.g. labour cost, expertise, availability, etc.) Although this paper focuses only on one factor which impacts task allocation in GSD projects (the distance factor), other factors could similarly be integrated within the SDaaS architecture in future works. The motivation for extending the SDaaS architecture is that it already support other aspects of GSD projects (as discussed in Sect. 2.1).

This paper comes as a first step towards a comprehensive approach for task allocation decision support within the SDaaS architecture. In the future, other factors affecting task allocation decisions need to integrated. It is also possible to adapt the model-based approach developed by Lamersdorf and Münch [13] which integrates three models: a risk model which identifies risks for each allocation alternative, an optimization model which suggests alternative allocation based on multiple criteria, and an effort overhead model which estimates the effort needed for each allocation alternative.

Acknowledgements B. Gallina is partially financially supported by EU and VINNOVA via the ECSEL Joint Undertaking under grant agreement No 692474, project name AMASS.

References

1. Software and Systems Process Engineering Meta-Model Specification, V2.0. formal/2008-04-01. Object Management Group (OMG), MA, USA (2008)
2. Alajrami, S., Gallina, B., Romanovsky, A.: Enabling global software development via cloud-based software process enactment. Technical Report TR-1494, Newcastle University, School of Computing Science (2016)
3. Alajrami, S., Gallina, B., Romanovsky, A.: EXE-SPEM: towards cloud-based executable software process models. In: MODELSWARD'16—Proceedings of the 4th International Conference on Model-Driven Engineering and Software Development, pp. 517–526. Scitepress, Rome, Italy 19–21 February (2016)
4. Alajrami, S., Gallina, B., Sljivo, I., Romanovsky, A., Isberg, P.: Towards cloud-based enactment of safety-related processes. In: Skavhaug, A., Guiochet, J., Bitsch, F. (eds.) Proceedings of Computer Safety, Reliability, and Security—35th International Conference, SAFECOMP'16, Trondheim, Norway, September 21–23, pp. 309–321. Springer (2016)
5. Alajrami, S., Romanovsky, A., Gallina, B.: Software development in the post-PC era: towards software development as a service. In: Abrahamsson, P., Jedlitschka, A. (eds.) The 17th International Conference on Product-Focused Software Process Improvement, PROFES'16, Trondheim, Norway, November 22–24, Proceedings. Springer (2016)

6. Carmel, E.: Global Software Teams: Collaborating Across Borders and Time Zones. Prentice Hall PTR, Upper Saddle River, NJ, USA (1999)
7. Carmel, E., Abbott, P.: Why 'nearshore' means that distance matters. Commun. ACM **50**(10), 40–46 (2007)
8. Conchúir, E.O., Ågerfalk, P., Olsson, H., Fitzgerald, B.: Global software development: where are the benefits?. Commun. ACM **52**(8), 127–131 (2009)
9. Grinter, R.E., Herbsleb, J.D., Perry, D.E.: The geography of coordination: dealing with distance in R&D work. In: Proceedings of the International ACM SIGGROUP Conference on Supporting Group Work. GROUP '99, pp. 306–315. ACM, New York, NY, USA (1999)
10. Hashmi, S.I., Clerc, V., Razavian, M., Manteli, C., Tamburri, D.A., Lago, P., Nitto, E.D., Richardson, I.: Using the cloud to facilitate global software development challenges. In: 2011 IEEE Sixth International Conference on Global Software Engineering Workshop, pp. 70–77 (2011)
11. Herbsleb, J.D., Moitra, D.: Global software development. IEEE Softw. **18**(2), 16–20 (2001)
12. Imtiaz, S., Ikram, N.: Dynamics of task allocation in global software development. J. Softw. Evol. Process **29**(1) (2017)
13. Lamersdorf, A., Münch, J.: Model-Based Task Allocation in Distributed Software Development, pp. 37–53. Springer, Berlin, Heidelberg (2010)
14. Lamersdorf, A., Munch, J., Rombach, D.: A survey on the state of the practice in distributed software development: criteria for task allocation. In: 2009 Fourth IEEE International Conference on Global Software Engineering, pp. 41–50 (2009)
15. Mockus, A., Weiss, D.M.: Globalization by chunking: a quantitative approach. IEEE Softw. **18**(2), 30–37 (2001). doi:10.1109/52.914737
16. Münch, J., Lamersdorf, A.: Systematic Task Allocation Evaluation in Distributed Software Development, pp. 228–237. Springer, Berlin, Heidelberg (2009)
17. Noll, J., Beecham, S.: Measuring global distance: a survey of distance factors and interventions, pp. 227–240. Springer (2016)
18. Setamanit, S.O., Wakeland, W., Raffo, D.: Planning and improving global software development process using simulation. In: Proceedings of the 2006 International Workshop on Global Software Development for the Practitioner, GSD '06, pp. 8–14. ACM, New York, NY, USA (2006)

Composite Event Handling over a Distributed Event-Based System

Amina Chaabane, Salma Bradai, Wassef Louati and Mohamed Jmaiel

Abstract The using of structured peer-to-peer networks improves system scalability but it confines users expressiveness in terms of desired exchanged data. To address this shortcoming, we exploit advantages offered by structured topology (Distributed Hash Table DHT) and extend it by novel approach in order to improve expressiveness by supporting Complex Event Processing (CEP). Our approach helps to make the right routing decision while avoiding the network overhead and preserving system scalability. It allows users to detail interest by defining logical and temporal patterns of exchanged data especially with the growth of data size encapsulated as events in the network. For efficient event filtering, we propose a smart data structure named CECube for rapid CEP over DHT. The CECube indexes firstly composite subscriptions, then basing on a simple binary search, it serves as publications filter and helps making the right decision for what events should be aggregated and forwarded to the adequate subscribers. The performance of our solution is implemented on Pastry DHT and evaluated using FreePastry simulator. The results demonstrate firstly that our approach is efficient in terms of filtering process and that the average number of routing nodes is decreased. Secondly, we prove the superiority of our approach as compared to another existing work.

A. Chaabane (✉)
Higher Institute of Applied Sciences and Technology, University of Kairouane,
B.P. 471, 1200 Kasserine, Tunisia
e-mail: amina.chaabane@redcad.org

S. Bradai
ReDCAD Laboratory, University of Sfax, National School of Engineers of Sfax,
B.P. 1173, 3038 Sfax, Tunisia
e-mail: Salma.bradai@redcad.org

W. Louati
Faculty of Economics and Management of Sfax, University of Sfax,
B.P. 1088, 3018 Sfax, Tunisia
e-mail: wassef.louati@redcad.org

M. Jmaiel
Research Center for Computer Science, Multimedia and Digital Data
Processing of Sfax, B.P. 275, 3021 Sfax, Tunisia
e-mail: mohamed.jmaiel@redcad.org

© Springer International Publishing AG 2018
R. Lee (ed.), *Software Engineering Research, Management and Applications*,
Studies in Computational Intelligence 722, DOI 10.1007/978-3-319-61388-8_12

1 Introduction

Fueled by the widespread adoption of new technologies and systems, such as the Machine to Machine (M2M) [1], gathering and sharing information become of great importance for the network's users. Given the example of crowd-powered sensor systems, that could provide highway air pollution information, surrounding noise level and more.

While current communication systems are using different protocols for message and information dissemination between scattered users, they are faced with the increased number of users, affect network efficiency and bothers users when they receive useless data.

As all kinds of information, from the internet to the cell phone, are driven by events, we thought of filtering exchanged data by integrating event-based systems [2, 3] as an effective messaging mechanism between components. From one hand, It offers the possibility to asynchronous data exchange in applications due to its specific characteristic of space, time, and flow decoupling. From the other hand, it allow users to interact with others and to specify the kind of messages or notifications they want to receive. Basing on an event service notification, event based systems ensure communication between scattered users. This event service is composed of brokers that are responsible for event filtering and routing. Its scalability depends also on the event service topology. It grows in importance from centralized to hierarchical and Peer-to-Peer (P2P) architectures. With P2P topology, event service topology can be structured or unstructured. Scalability is more sophisticated with structured topology based on Distributed Hash Tables such as Pastry and Chord.

However, messages come with a high rate but few of them satisfy user's interests. In fact, they still be sent unnecessarily and passed across our systems as unrelated pieces of information. Moreover, users have more and more various events for publication and attempt to search for multifarious events. With information diversity, expressiveness of these systems is a salient feature to evaluate system efficiency against user's requirements. Publish/Subscribe (Pub/Sub) system handles expressiveness mainly with topic-based and content-based systems. But, user can require or produce many events concurrently or sequentially with various contents and topics. User's interest becomes more exigent when he searches to reorganize favorite events reception or detect significant event or event resulting from occurrence of other events. Consequently, a lot of events could come with a high rate, but few of them could satisfy user's interests.

This issue can be handled through Complex Event Processing (CEP) approaches. In fact, CEP offers the possibilities to aggregate and correlate events together by a set of understood relationships in order to make them a source of great power, called composite events, that can yield a wealth of information. Those composite events detail events composition with temporal and logical relationships and even spatial relationship for sensor networks especially. Let us consider one example to illustrate the problem involved in case of composite event handling. Taking the example with centralized event service shown in Fig. 1, at the left hand we present coming

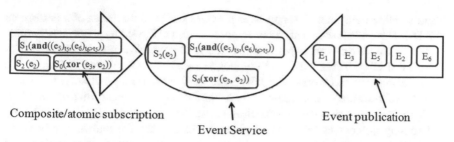

Fig. 1 Composite event handling

subscription, at the right coming publication and at the bottom, we found the event service responsible to event storage, matching and routing. In this example, S_0 link up e_2 and e_3 with *XOR* logical relationship. Thus, if the event service receives E_3 followed by E_2 matching e_3, e_2 respectively, then, it must memorize the first one (E_3) to be notified and ignore E_2. However, E_2 matches S_2, consequently it will be sent to the corresponding subscriber(s). When handling S_1, the event service should check temporal and logical relationships to build the composite event. For example, when receiving E_6, the event service should check the receiving of E_5 previously in order to validate temporal relationship. Otherwise, clients will be overwhelmed by an excessive amount of primitive events, most of which may be irrelevant and could be filtered out before reaching the client. Thus composite event allows to better satisfy user's interest and reduces network traffic by getting rid of uninterested events from the network.

However, collecting events from scattered brokers and aggregating them to composite events according to different event patterns, brings new challenges in terms of their routing and management. Widely explored in the literature, this issue stills require improvement for distributed event service, in particular event service under P2P architecture for CEP.

After having studied existing approaches, this paper relies on Pub/Sub system based on Distributed Hash Tables (DHT), which intrinsically offers efficient functionalities such as event routing flexibility, scalability, load balancing and fault tolerance, in order to manage distributed composite events routing. We use topics to compute keys for routing atomic/composite events in a DHT. Our Pub/Sub communication layer will be responsible for disseminating composite topic into primitive and sub-composite topics that are mapped on nodes responsible on their hashed keys. Furthermore the paper concentrates on combining all logical and temporal relationships on a scalable P2P event-based system. We aim to provide a composite event management solution that deals with all composite event patterns and structured P2P networks. To this end, we propose a three dimensional indexing hash space named CECube for detection of composite events produced throughout a distributed network. While basing on binary search, it allows not only an efficient events matching and filtering process, but also it reduces useless transfer of atomic/primitive events throughout brokers' network. This is by checking temporal and logical constraints before their sent. Our approach provides CEP requirements of a performed P2P Pub/Sub middleware that can be integrated with IoT systems or social networks.

The remainder of this paper is organized as follows. We remind the notion of Distributed Hash Table, CEP and composite event as backgrounds in Sect. 2. We detail existing related works with advantages and shortcomings in Sect. 3. Then in Sect. 4, we introduce our proposed approach of composite event filtering based on DHT to overcome previous works limitations. Thereafter, we detail the indexing and routing process in subscription phase in Sect. 5 and the matching process in publication phase in Sect. 6. Results obtained from several experiments are provided in Sect. 7. We finish this paper with a conclusion and future work in Sect. 8.

2 Backgrounds

To better understand our approach, we detail all its pillars in this section.

2.1 Distributed Hash Table Overview

DHT provides a lookup service similar to a hash table. The (Key, Value) pairs are stored in a DHT node (rendezvous node). Each DHT node has a unique identifier (nodeID) and stores some (key, value) pairs which have closest keys to its nodeID. Generally, nodes are organized virtually according to the order growing of nodes identifiers on a DHT ring (see Pastry [4], Chord [5], etc.).

2.2 Complex Event Processing

CEP can be seen as an extension to traditional Pub/Sub system, which allows subscribers to express their interest in composite events. It consists in collecting information produced by multiple, distributed sources, to process it in a timely way, in order to extract new knowledge or valuable event as soon as the relevant information is collected. It is based on composite event patterns which define knowledge and desired event composition.

2.3 Composite Event

We can consider the following definition of event composition as proposed in [6]: "Composite subscriptions consist of atomic subscriptions linked by logical or temporal operators, and can be used to express interest in composite events. A composite subscription is matched only after all component atomic subscriptions are satisfied".

2.4 Composite Event Relationships

In event-based system, user's interest can be conveyed by a primitive event (simple) or composite event when there is a dependency or relativity between primitive events. This dependency can be represented by the composite event relationships. We handle logical and temporal relationships in our work.

2.4.1 Logical Relationships

The logical relationships supported by our approach are:

- **Conjunction (And)**: Events E_1 and E_2 should occur;
- **Disjunction (Or)**: Event E_1 or E_2 should occur;
- **Or-exclusive (XOR)**: Event E_1 or E_2 should only one occurs;
- **Negation (Not)**: Event E should not occur.

2.4.2 Temporal Relationships

In addition to logical operators, we handle also the following temporal relationships:

- E_1 **before** E_2: When user requires E_2 after finishing E_1;
- E_1 **meets** E_2: When user requires receiving E_1 and E_2 with time intersection;
- E_1 **overlaps** E_2: When user requires receiving E_1 and E_2 with time intersection but E_1 starts the first and E_2 finishes the last;
- E_1 **finishes** E_2: When user requires E_1 and E_2 that finishes in the same time;
- E_1 **includes** E_2: When user requires to receive E_2 when receiving E_1;
- E_1 **starts** E_2: E_1 and E_2 start in the same time;
- E_1 **equals** E_2: E_1 and E_2 start and finish in the same times.

3 Related Work

In this section, we provide an overview of several research efforts in the literature focused on the CEP. Then we discuss Pub/Sub systems and CEP applied to social networks, IoT and crowdsensing systems.

3.1 CEP

The CEP appears as an important issue in the Pub/Sub systems due to user requirements value. In fact, taking account of composite user interest provides a flexible and energy-efficient manner and performs near real-time processing of Big Data streams. Existing solutions can be categorized into centralized and distributed according to the event service topology. The centralized solutions appear first and are more efficient than distributed solutions for complex event routing in terms of matching rapidity and event expressiveness. RUBCES [7] is a Rule Based Composite Event System defining a Storage and Management of subscriptions as a centralized entity based on defined rules for composite event matching and management. RUBCES allows achievement and detection of primitive and composite events with only logical relationships (and, or). However, its centralized architecture causes server crowding and reduces its reliability since the server can fail and as a result all subscriptions and publications fail to.

Pietzuch et al. [8] propose a Composite Event Detectors (CED) based on a core CE language compiled on Finite State Automata (FSA). These CED are devoted to detect concurrent composite event patterns with specific parameterization and a rich time mode. However, they cannot be deployed on distributed network. While CED suffer especially from scalability issue with centralized automata detectors, authors propose to improve this solution by duplicating these detectors at favourable locations according to the network bandwidth and sources of composite events [8]. Nevertheless, this proposition remains suffer from scalability problem. In fact, when some composite events are required too much, nodes of CED would be certainly overcrowded. It suffers also from user-friendly definition of new composite event according to user requirements as he needs to define a new CED.

Others works appear to purge scalability shortcoming and enhance composite event relationships. In this context, Courtenage et al. [9] propose Composite Event Detectors and Atomic Event Detectors created on different nodes according to received subscriptions. The identity of an event detector broker is located in the network by hashing the event type and route it to a node having the closest successor identifier to the service identifier (AED/CED). The event definition is based on the λ-calculus formal language. Consequently, this solution lacks support for temporal relationships which are a cornerstone for CE expressiveness and usefulness.

Lai et al. [10] handle composite event detection for sensor networks with temporal, logical and spatial event composition. Each sensor node is programmed to detect specific composite event according to program images conveying the composite event and the sensor nodes responsibilities. Thus, the definition of new composite event requires new program definition which is out of the simple user scope. Besides, this approach can be used only for sensor networks deployed on limited area such as on an airport to control freight and passenger traffic or on an enterprise to control temperature versus working hours. However, the deployment of our framework is possible on wide area network and it allows users to define new subscriptions according to their interests.

Last work is named JTangCSPS proposed by Qian et al. [11]. It is a composite and semantic Pub/Sub system over structured P2P networks. They use OWL language to describe semantic events. Since OWL does not allow defining logical and temporal operators, they use RDF graphs to describe the relations of composite event. The major shortcoming of this work is that all primitive events shall be reached and checked before the checking of the composite event relationships which is unnecessary in some cases as when using the OR and XOR operators.

There are few approaches that have succeeded to maintain CEs management, but some of them fail to comply with the event expressiveness such as the breach of some logical and temporal relationships, and others are built on unstructured or centralized event notification system. In this paper, we present a composite event pattern modeling and distributed composite event management to achieve large-scale and efficient routing over P2P Pub/Sub system. In the next, we discuss CEP application to the IoT, Crowd sensing and social network.

3.2 CEP Could Be Applied to IoT, Crowd Sensing Systems and Social Networks

IoT and mobile Crowd Sensing are responsible to supervise and collect data over a large network of senors and mobile devices which results high traffic. Recently, few approaches propose to use CEP to detect valuable events for real time IoT application. Chen et al. propose a distributed CEP engine for IoT applications [12]. They use only logical operators to define complex event patterns. The CenceMe application retrieves and publish automatically sensing people's presence to social networks through mobile phones [13]. It generates a lot of traffic by sensing data, but without any filtering on generating data. It uses the phones and the backend servers to achieve scalable inference. In the same context, PEIR [14] is also a participatory sensing application based on GPS location data collection using mobile phone. It estimates personal exposure to pollution and environmental impact. It uses client-server architecture which affects scalability of the system. It is also not real time application and does not filter exchanged data.

New mobile sensing platforms using data filtering are recently proposed in order to minimize network traffic. Lifestreams [15] is a modular sense-making tool-set for identifying important patterns from everyday life. It is based on data analysis software. This software collects data from mobile phone into centralized data base to be analyzed to facilitate the exploration and evaluation of personal data stream sense-making. Lifestream uses some defined views to analyze collected data and identify key behaviors and trends which are relevant to an individual's health as well as to enable researchers from health domains to identify behavioral-indicators from large volumes of raw and heterogeneous data streams according defined view.

Some IoT applications need real time processing for handling big data streams. They use Pub/Sub system to collect interested data by simple event filtering. We find Pogo middleware [16], an application used on mobile phones to facilitate the access to sensor data for the research community, uses Pub/Sub system with simple topic-based for filtering. It aims to achieve energy-saving on mobile devices by simple filtering of sensed data on mobile devices. Tong et al. [17] propose an ubiquitous Pub/Sub Platform for wireless sensor networks. They provide content-based Pub/Sub with high level of abstraction from the underlying sensors and network infrastructures. Users can subscribe for sensing data by simply specifying the target area, sensing types and data ranges of interest. CUPUS [18] for CloUd-based PUblish/Subscribe middleware, is a mobile crowd sensing system that reduces energy consumption significantly on mobile devices and sensors by suppressing the transmission of redundant and irrelevant data into the cloud. To summarize, there are few IoT applications using Pub/Sub system with simple filtering which reduce energy consumption and traffic. These objectives can be more satisfactory with CEP. Especially for the social networks that attract a majority of the Internet users which increase significantly User Generated Content [19]. Unfortunately, users are not always satisfied by received contents so the checking of composite user interests reduces the network traffic and improve the network efficiency.

In the next, we detail our CEP approach for event filtering over DHT.

4 The Proposed Approach of Composite Event Filtering Based on DHT

4.1 Complex Event Modeling

In our work, we define a complex event based on a composite event definition as an aggregation of primitive or sub-composite events with a set of logical operators and temporal constraints. Therefore, We formulate a Composite Event (CE) made up of the set of events E, where E can be a primitive or a sub-composite event with as follows:

$$CE = [op_i(E_{i_{/Ti}}, E_{j_{/Tj}})] \tag{1}$$

With:

- E_i/E_j: primitif or sub-composite event; so that a CE could be the aggregation of two sub-composite events, two primitive events or the aggregation of a primitive event and a sub-composite event;
- T_i/T_j: temporal constraint of E_i/E_j;
- op_i: logical/temporal (Log/Temp) relationship between E_i and E_j.

For example to compose three primitive events with logical operator "and", according to our formula, the generated CE will be as follows: CE = [and $(E_{3_{/T1}})$, [and $(E_{1_{/T1}}, E_{2_{/T2}})$]].

4.2 Approach Overview

In our approach, we extend structured P2P Pub/Sub system for topic-based event in order to support composite topic-based event while relying on DHT protocol. The specific contributions of our work revolve around three main pillars.

Fig. 2 Decomposition tree of CE_1

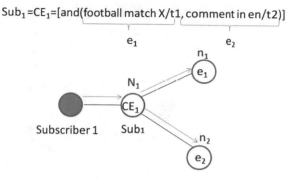

$$Sub_1 = CE_1 = [and(football\ match\ X/t1, comment\ in\ en/t2)]$$

Firstly, brokers could act either as event sources and event consumers. They act also as rendez-vous/filter nodes between publishers and subscribers. In fact, matching publications into subscriptions needs that subscriptions and publications for particular events meet at a certain nodes in the system where they can be compared.

Secondly, In case of a composite subscription (CS) (subscription with a composite event), our Pub/Sub communication layer is responsible on its decomposing into primitive and sub-composite events called its members. Those members are mapped later to their root nodes responsible on their hashed identifiers. The decomposition and mapping process follows a tree structure shown by Fig. 2. The example in the figure shows that subscriber1 desires the reception of CE_1 which is is the aggregation by the logical operator "and" of two primitives events e_1: foot ball match and e_2: comments in english. According to this example, the tree structure is built in subscription phase from parent (N_1) to roots $(n_1\ and\ n_2)$. In publication phase, it is followed inversely so that the event composition will be checked gradually on rendez-vous nodes. Useless events that have not meet any primitive and/or composite subscription stop their dissemination over the tree, and hence reducing the network traffic.

Moreover, our proposed decomposition process can detect shared or repeated primitive or composite events. As shown the Fig. 3, the composite subscription Sub_1 is reused in the composite subscription sub_2. At this stage, the re-decomposition of sub_1 and re-routing of its members is unnecessary as it was already performed when handling sub_2. Consequently, we avoid again unnecessary events transfer.

Thirdly, we propose two smart structures, the "plane" and "CECube" for indexing the primitive and composite subscriptions respectively. For both of them, the indexing process is based on an EventFlag $\in \{0, 1\}$ to indicate the subscription existance in the current rooting node. Our objective behind using those structures is to make matching events easier and more efficient in the same time. Regarding the "Plane" structure, its main role is to check logical and temporal constraints of a primitive subscription.

Regarding the "CECube", it allows CS indexing according to its members. More details of the "plane" and "CECube" indexing and matching processes are given in Sects. 5 and 6 respectively.

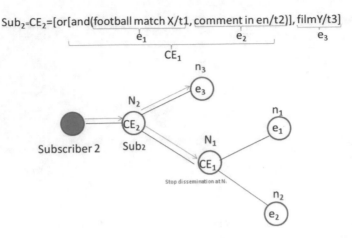

Fig. 3 Decomposition tree with CE_1 as a common event

5 Primitive/Composite Event Indexing in Subscription Phase

5.1 Indexing Primitive Subscription Through Plane Structure

To store primitive events on a responsible broker, we define a Plane structure with two axes $I(e)$ and $I(T)$ as shown in Fig. 4, where $I(e)$ is the hashed value index of primitive event and $I(T)$ is the hashed value index of the event occurrence time. We use the uniform hashing function sha-1 that can accommodate all IDs without conflict.

Each cell in this Plane contains a value denoted as PlaneValue. It consists of two pieces of information. The first is the EventFlag, which is a bit value that indicates whether a subscription contains a primitive event "e" at time T. Therefore, the Event-Flag is set to "1" in order to indicate that there is a subscription for an event e_i at time T_i. The second is a Subscriber Identifier Vector (SIV) that stores identifiers of

Fig. 4 Plan structure

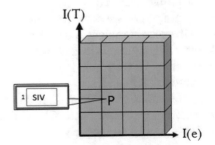

subscribers to the primitive event. Subscribers could be either nodes responsible on composite or simple subscriptions. In fact, primitive events can match primitive subscriptions or belong to one or more composite subscriptions, to be aggregated later (using the cube structure) with other members of the same CE.

5.2 Indexing Composite Subscription Through CECube and Composite Event Matching Vector (CEMV)

To provide efficient matching operations in a distributed event service, we propose a three dimensional indexing hash space named CECube. The CECube is maintained by each broker that is responsible for a composite subscription. It is used to store composite topic pattern, i.e. its aggregated members, and its logical and temporal relationships. The axes respectively represent the hashed value index for composite event CE, primitive or sub-CE belonging to CE denoted E and the occurrence time T. The axes indexing process is performed similarly to the Plane axes. We note that sub-CE or CE hashing is performed without event relationships and events time, which reduces the number of built cube. The occurence time of sub-CE determines the last occurence time of its events E.

When a broker receives a composite subscription, it begins by defining its members (primitive events and/or sub-CEs) in addition to their Log/Temp relationships. Then, it uses its cube structure in order to index and map all members into their corresponding cells.

As shown in Fig. 5, a Plane perpendicular to the I(CE) axis is identified as [I(CE), *, *] and denoted as "CellMatrix". It represents the composition of the corresponding CE at any time. Similarly, a line parallel to the I(E) axis forms the composition of CE at a specific time T. It represents the CellSequence and is identified as [I(CE), *, I(T)]. Each cell of the cube [I(CE), I(E), I(T)] is denoted as cubeValueSet and maintains a set of values as follows:

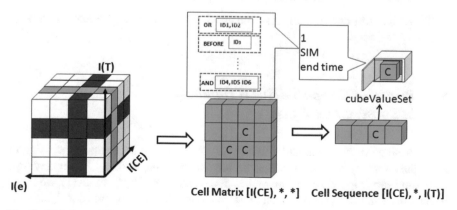

Fig. 5 CECube structure

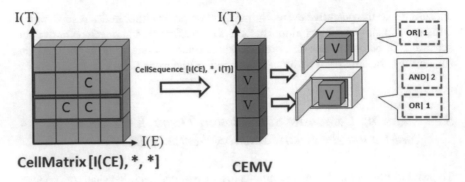

Fig. 6 Composite Event Matching Vector modeling

- The EventFlag $\in \{0, 1\}$: as a bit that indicates whether a composite subscription with (CE) contains an event E (primitive or composite) at time T.
- Subscriber Identifier Map denoted as SIM: It contains pairs of (key, value) as follows:

  - The key is the relationship;
  - the value is the the corresponding vector of subscribers' identifiers (SIV). In other words, they are subscribers to the concerned CE aggregated with the same relationship.

- The end time of the event/sub-CE: necessarily for temporal relationships checking.

To check event composition total matching efficiently, we define a *Composite Event Matching Vector* (CEMV) for each CE. Our aim is to check easily the total matching of a given CE with the coming of its members. As shown the Fig. 6, values of each element in the CEMV will contain pairs of (key, value) as follows:

- The key: is relationship
- The value: is the EventFlag summation in the CellSequence [I(CE), *, I(T)] of the concerned CE aggregated with the same relationship indicated in the key.

Those two information will make the matching easier in the publication phase as explained in Sect. 6.

5.3 Primitive/Composite Event Handling in Subscription Phase

While basing on the Plane and CECube structures, the subscription processing of a composite and primitive subscriptions are shown in Algorithm 1. On receiving a composite subscription according to the CE pattern defined by formula (1), we check if the same composite subscription (Sub) is already indexed before (line 3). In case that the concerned routing node has never received similar composition of Sub, it starts by decomposing the subscription to reveal its members (E). Then, According to the hash indexes of CE and E, their start time, it indexes those members in their corresponding cells in the cube. Therefore, each cell of [I(Sub), I(E), I(T)] will contain the EventFlag set to 1, the SIM containing pairs of [[relationship, Vector of subscriber's ID (SIV)] and the end time of the event (E). Then the concerned Composite Event Matching Vector is updated with adding the pair [relationship, summation of EventFlag]. Then, the events (E) will be routed to their rooting nodes, that will process the same in case that (E) is itself a composition (lines 4–8).

If Sub is already indexed before, the process of cube indexing will just check the relationship. If it is the same as the already indexed subscription, it will just add the subscribers' identifier to the adequate pair in the SIM. Otherwise, it adds an other pair [relationship, SIV] in the same SIM. Then, the decomposition is stopped as events subscription are already routed by the alercady indexed subscription (lines 10–11).

When a rendezvous node receives a primitive subscription (line 15), it updates its Plane structure by setting corresponding EventFlag to "1" and adding sender node to the SIV using *plane* function (line 16).

Algorithm 1: Subscription processing

Data: subscription Sub, sender S
Result: E1, E2, op
1 switch *type of Sub* do
2 | case *Composite*
3 | | if *Sub never mapped* then
4 | | | *decompose(Sub, E1, E2, op)*;
5 | | | *cube(Sub, E1, E2, op, S)*;
6 | | | *updateCompositeEventMatchingVector (Sub, E1, E2)*;
7 | | | *send(E1)*;
8 | | | *send(E2)*;
9 | | end
10 | | else
11 | | | *updateSIM(S)*;
12 | | end
13 | | break;
14 | endsw
15 | case *Primitive*
16 | | *plane(Sub, S)*;
17 | | break;
18 | endsw
19 endsw

Figure 7 depicts the example of composite subscription shown previously by Fig. 3. For composite subscription sub_2, a cube structure is created in N_2 as the root node of the CE_2. Then, members of CE_2 are mapped in the cube according to their hash index of events and occurence time. The occurence time of the sub-CE (CE_1) is the last occurence time of its members e_1 and e_2 (they have the same occurence time

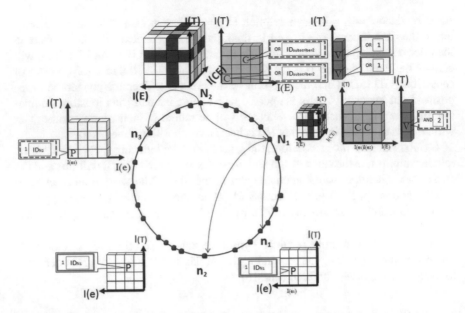

Fig. 7 Composite subscription processing

in the example as the match should coincide with comments). The Composite Event Matching Vector responsible on the Plane $[I(CE_2), *, *]$ is updated by adding the pair [logical relationship ("OR" in the example), summation of the EventFlag = 1] for the CellSequence $[I(CE_2), *, I(t)]$. Then event members are routed to their root nodes (N_1 and n_3 respectively). The same scenario is performed in N_1 responsible on CE_1.

When primitive events are routed to their root nodes, a Plane structure is created to index those events according to their ID and time occurence. For example, for e_1 the concerned PlaneValue will contain the EventFlag set to "1" and the SIV containing the ID of its subscriber which is N_1 in the example.

6 Matching of Primitive/Composite Event in Publication Phase

The publication phase consists in collecting and aggregating all primitive and/or composite topics from different brokers to satisfy composite subscriptions. The aggregation process will respect tree decomposition from leafs to the root in order to avoid unnecessary event notification when primitive event does not satisfy logical and temporal constraints.

The Algorithm 2 details steps of the publication phase. We distinguish primitive event from composite event handling. In the first case, a primitive publication is received on the root node of primitive event (line 1). We check through the Plane created in the subscription phase if the corresponding EventFlag is set to "1" (line 2).

Then, we use SIV of the PlaneValue for sending primitive events to their subscribers (lines 3–5).

In the second case, an event E (primitive or sub-CE) is received by a node that maintains a cube structure (line 7). We look at the Plane [*, I(E), *] perpendicular to the I(E) axis to check if it belongs to any composite subscription (line 8). When EventFlag is set to "1", we look up the corresponding [I(CE), *, *] CellMatrixs and scan all cubeValueSets for event matching and relationships checking and to found following destinations (SIV) (lines 9, 10).

As the event E is reached, the EventFlag in the concerned cell which was already set to 1 in the indexing process will be set to 0 in this matching process (line 11). Note that we take into account the relationship when updating the EventFlag. For example for the logical relationship "OR", once an event is reached, we update also the EventFlag of the other member and set it to 0. After verifying Log/Tem relationship with other events (line 12), we update the CEMV to check the total matching of the concerned CE (line 13). As the CEMV cells contain pairs of [Log/Temp relationship, summation of the EventFlag] and EventFlag of reached events are set to 0, when all summations of the EventFlag for the same relationship are set to 0, we reveal that the concerned CE is matched. In this stage, a notification is sent to the subscriber using the adequate SIV in the adequate SIM stored in the cubeValueSet. Finally, we send composite event to those subscribers (lines 14-16-17).

Algorithm 2: Publication processing

Data: publication e, receiver R
Result: E1, E2, op

```
1  if R_has_a_plane_structure and e_is_primitive then
2  |    SIV ⟵ scanPlan(e);
3  |    for dest ∈ SIV do
4  |    |    send(e, dest);
5  |    end
6  end
7  if R_has_Cube_structure then
8  |    CE ⟵ scanPlan([*, I(E), *]);
9  |    for ce ∈ CE do
10 |    |    if scanCubeValue([I(CE), *, *]) then
11 |    |    |    updateEventFlag;
12 |    |    |    checkLog/TemRelationship;
13 |    |    |    updateCompositeEventMatchingVector(ce);
14 |    |    |    if (CompositeEventMatchingVector(ce)==0) then
15 |    |    |    |    SIV ⟵ scanPlan(ce);
16 |    |    |    |    for dest ∈ SIV do
17 |    |    |    |    |    send(ce, dest);
18 |    |    |    |    end
19 |    |    |    end
20 |    |    end
21 |    end
22 end
```

7 Experimental Results

Experimental results consist in two parts. The first part provides an experimental evaluation and explanation of the benefits of our solution. In the second part, we compare it to JTangCSPS system to check its performance in terms of routing delay.

We have implemented our system over Scribe Pub/Sub system based on Pastry DHT. Our evaluation uses FreePastry simulator with almost the same conditions of JTangCSPS. Our measurements take place also on a standard PC installation with Linux libraries and a hardware configuration comprising Intel core i5 CPU 2.53 GHz, 4 GB RAM. The experimentations are made with a static DHT network that does not suffer from node failure.

7.1 Network Traffic Reduction

To achieve our goal, we propose a CECube structure for event composition through P2P Pub/Sub system. Our approach allows to meet the needs of users and consequently reduce the network traffic by avoiding unnecessary transfers between users.

We propose two scenarios to evaluate our approach as far as traffic reduction is concerned. In the first scenario, we inject 200 composite subscriptions, each one is composed of five primitives subscriptions. This scenario allows to compare a Pub/Sub system with event composition over existing communication networks. In the second scenario, we inject the 1000 atomic subscriptions that make up the 200 composite subscriptions of the first scenario. This scenario helps to show the contribution of event composition with distributed Pub/Sub system. We measure the number of notifications received by customers in the different scenarios after 4000 publications sent over the network.

The result shows that the number of notifications sent to the customer is greatly reduced by using an event-based system. Indeed, the flow of publications can reach 100% if we consider the case of Facebook where sharing is done between a group of friends. This type of sharing does not consider the interests of customers and as result we observe user frustration. However, considering the interests of customers, the reduction can reach 82%, as shown by the curve. Moreover, the injection of composite subscriptions reduced more the overall traffic. The reduction reaches 65% compared to a system using Pub/Sub without composition (first scenario). These results are shown by curves of Fig. 8.

7.2 Evaluation of Our Approach for Composite Subscriptions with and Without Intersections

We note that in the practice, there is too much intersection between user interests above all when some events are famous and shared between most of the population.

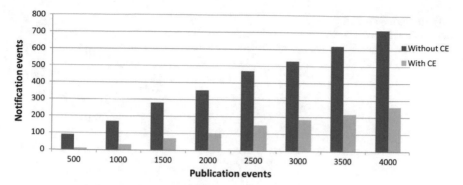

Fig. 8 Traffic reduction by Pub/Sub system with event composition

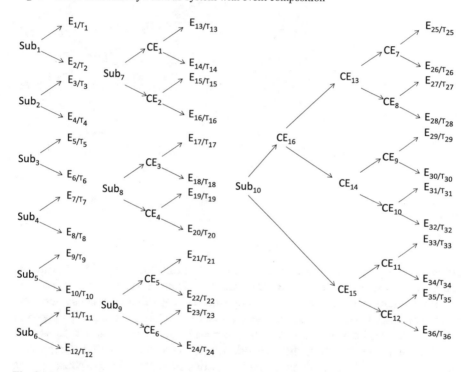

Fig. 9 Example of 10 composite subscriptions without intersections

The proposed CECube structure makes the intersection detection easier and allows checking the matching of several events simultaneously. To check CECube performance, we propose to compare the routing delay and hops count with two patterns of composite events as shown by Figs. 9 and 10.

The first one is without intersection between composite subscriptions and it is made up of 36 primitive subscriptions composing 10 composite subscriptions (Fig. 9). For the second one shown by Fig. 10, each composite subscription sent is

Fig. 10 Example of 10
composite subscriptions with
intersections

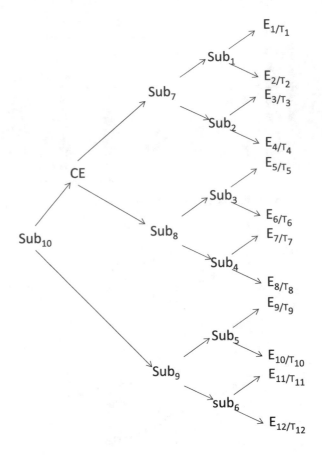

used again as a sub-composite subscription in the following one. So 12 primitive subscriptions allow to produce 10 composite subscriptions sent to the event service.

For evaluation, we measure the routing delay and hops when increasing networks size. The routing delay of a composite subscription includes the network delay and the time to build cubes and to split the composite subscription at each node. For publication phase, it includes the network delay and the matching time. Comparing the two curves of Fig. 11, we note that the routing time of composite subscriptions without intersection is slightly higher compared to the one of composite subscriptions with intersections. The difference is 150 ms for the simple example shown in Fig. 10 but it can be more significant certainly in the practice when having much intersection between user interests. Indeed, the number of created cube is lowered when the compositions are repeated even with different composition relationships. We also remember that the hash of composite event is applied on the concatenation of primitive events without including operators and relationships. Therefore, when primitive or sub-composite subscriptions are repeated even with different relationships, the matching of the composite or sub-composite events is carried out in the

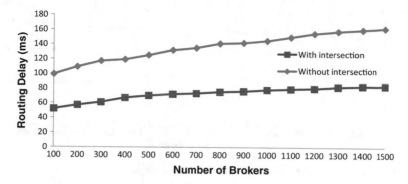

Fig. 11 Routing delay versus event service size in subscription phase

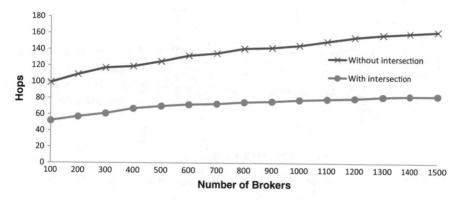

Fig. 12 Hops account versus number of nodes in subscription phase

same cube. The number of visited nodes is also reduced since a responsible node to a composition is already found, the subscription stops without finishing its decomposition and without routing of these primitive subscriptions.

This analysis is also verified by experimental results shown by the curves of Fig. 12. These curves show the hops count for composite subscription routing with and without event repetitions. The number of hops is significantly reduced with subscriptions made by repetition with the patterns of Fig. 10. The difference is on average 40 hops with this example and can be more important in practice by increasing the number of DHT nodes and varying user interests.

Now, we compare the routing delay in the publication phase with the same patterns shown previously by Fig. 10. The curve of routing delay of publication when having CE with intersection is almost constant when increasing network size. This proves the scalability of our system Fig. 13.

Comparing two curves of this figure, we note the important difference in routing delay between CE with and without repetition. When primitive or sub-composite subscriptions are repeated even with different relationships, the event matching is carried out in the same cube.

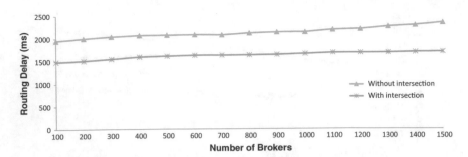

Fig. 13 Routing delay versus number of nodes in publication phase

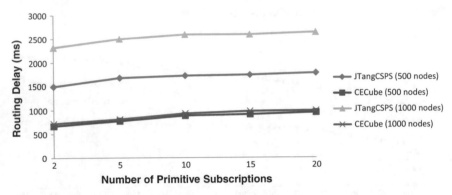

Fig. 14 Routing delay versus number of primitive subscriptions with JtangCSPS and CECube

Finally, we compare our approach against JTangCSPS, explained in Sect. 3 in terms of routing delay in order to evaluate our system performance. The routing delay of a composite subscription includes the network delay and the time to build the composite subscription tree and cube in JtangCSPS and CECube respectively and to split the composite subscription at each node. As JtangCSPS evaluation, we have tested with different scenarios when the number of primitive subscriptions of the composite subscription varies from 2 to 20. Each broker randomly makes 20/i composite subscriptions, each consisting of i primitive subscriptions where $i = 2, 5, 10$ and 20. To simplify the comparison, only operator conjunction "AND" is used in composite subscription, and each primitive subscription only contains one type constraint.

Figure 14 shows that CECube has a lower routing delay than JTangCSPS. This is because, in CECube, index construction in the cube architecture is less expensive than the tree construction with JTangCSPS. With JTangCSPS, the routing delay increases fast before NoPS reaches 5 and then it increases slowly. It is the same aspect with our CECube. Scalability of our system is explained by the stability of routing delay which is almost the same with CECube regardless of the number of nodes. However, it increases by 800 ms between 500 and 1000 nodes with JTangCSPS.

8 Conclusion and Future Work

In this paper, we present a new three dimensional hash space, the CECube to perform CEP distributed over structured P2P network. Our approach proposes a plane and cube structures. The plane structure is used to index primitive event according to time occurrence and event ID hashing with bit indication. It simplifies event occurrence checking. The CEcube structure indexes composite events based on a binary information in the subscription phase. In publication phase, matching process is performed while basing on this simple binary information. This approach reduces the required amount of events that has to be transferred on the network. With experimental results, we demonstrate that our approach reduces significantly the routing delay comparing to JtangCSPS system and it stills almost the same with increased network size.

An interesting direction of future research therefore would be to enhance semantic aspect of our approach in order to fulfill user interests. It is also interesting to exploit our approach in various application domains such as the social network and IoT.

References

1. Alaya, M.B., Banouar, Y., Monteil, T., Chassot, C., Drira, K.: Om2m: extensible etsi-compliant M2M service platform with self-configuration capability. Proc. Comput. Sci. **32**, 1079–1086 (2014). In: The 5th International Conference on Ambient Systems, Networks and Technologies (ANT-2014), the 4th International Conference on Sustainable Energy Information Technology (SEIT-2014) [Online]. http://www.sciencedirect.com/science/article/pii/S1877050914007364
2. Carzaniga, A., Rosenblum, D.S., Wolf, A.L.: Design and evaluation of a wide-area event notification service. ACM Trans. Inf. Syst. Secur. (TISSEC) **19**(3), 332–383 (2001)
3. Liu, Y., Plale, B.: Survey of publish subscribe event systems (2003)
4. Rowstron, A.I.T., Druschel, P.: Pastry: scalable, decentralized object location, and routing for large-scale peer-to-peer systems. In: Proceedings of the IFIP/ACM International Conference on Distributed Systems Platforms Heidelberg, pp. 329–350. Springer (2001)
5. Stoica, I., Morris, R., Karger, D., Kaashoek, M.F., Balakrishnan, H.: Chord: a scalable peer-to-peer lookup service for internet applications. In: Proceedings of the 2001 Conference on Applications, Technologies, Architectures, and Protocols for Computer Communications, pp. 149–160. ACM (2001)
6. Hinze, A., Buchmann, A.: Principles and Applications of Distributed Event-Based Systems. IGI Global (2010)
7. Sahingoz, O.K., Erdogan, N.: Rubces: rule based. composite event system. In: XII. Turkish Artificial Intelligence and Neural Network Symposium (TAINN), Turkey (2003)
8. Pietzuch, P.R., Shand, B., Bacon, J.: A framework for event composition in distributed systems. In: Proceedings of the ACM/IFIP/USENIX 2003 International Conference on Middleware, ser. Middleware '03, pp. 62–82. Springer, New York, Inc. (2003)
9. Courtenage, S., Williams, S.: The design and implementation of a p2p-based composite event notification system. In: Proceedings of the 20th International Conference on Advanced Information Networking and Applications - Volume 01, ser. AINA '06, pp. 701–706. IEEE Computer Society (2006)
10. Lai, S., Cao, J., Zheng, Y.: Psware: a publish/subscribe middleware supporting composite event in wireless sensor network. In: Seventh Annual IEEE International Conference on Pervasive

Computing and Communications—Workshops (PerCom Workshops), Galveston, TX, USA, pp. 1–6 (2009)

11. Qian, J., Yin, J., Dong, J., Shi, D.: Jtangcsps: a composite and semantic publish/subscribe system over structured p2p networks. Eng. Appl. Artif. Intell. **24**(8), 1487–1498 (2011)

12. Chen, C., Fu, J.H., Sung, T., Wang, P., Jou, E., Feng, M.: Complex event processing for the internet of things and its applications. In: 2014 IEEE International Conference on Automation Science and Engineering, CASE 2014, New Taipei, Taiwan, 18–22 August, 2014, pp. 1144–1149 (2014)

13. Miluzzo, E., Lane, N.D., Fodor, K., Peterson, R., Lu, H., Musolesi, M., Eisenman, S.B., Zheng, X., Campbell, A.T.: Sensing meets mobile social networks: The design, implementation and evaluation of the cenceme application. In: Proceedings of the 6th ACM Conference on Embedded Network Sensor Systems, ser. SenSys '08, pp. 337–350. ACM (2008)

14. Mun, M., Reddy, S., Shilton, K., Yau, N., Burke, J., Estrin, D., Hansen, M., Howard, E., West, R., Boda, P.: Peir, the personal environmental impact report, as a platform for participatory sensing systems research. In: Proceedings of the 7th International Conference on Mobile Systems, Applications, and Services, ser. MobiSys '09, pp. 55–68. ACM (2009)

15. Hsieh, C., Tangmunarunkit, H., Alquaddoomi, F., Jenkins, J., Kang, J., Ketcham, C., Longstaff, B., Selsky, J., Dawson, B., Swendeman, D., Estrin, D., Ramanathan, N.: Lifestreams: a modular sense-making toolset for identifying important patterns from everyday life. In: The 11th ACM Conference on Embedded Network Sensor Systems, SenSys '13, Roma, Italy, 11-15 November, 2013, pp. 5:1–5:13 (2013)

16. Brouwers, N., Langendoen, K.: Pogo, a middleware for mobile phone sensing. In: Proceedings of the 13th International Middleware Conference, ser. Middleware '12, pp. 21–40. Springer, New York, Inc. (2012)

17. Tong, X., Ngai, E.C.H.: A ubiquitous publish/subscribe platform for wireless sensor networks with mobile mules. In: IEEE 8th International Conference on Distributed Computing in Sensor Systems, DCOSS 2012, Hangzhou, China, pp. 99–108 (2012)

18. Antonic, A., Marjanovic, M., Pripuzic, K., Podnar Zarko, I.: A mobile crowd sensing ecosystem enabled by cupus. Future Gener. Comput. Syst. **56**(C), 607–622 (2016)

19. Paper, C.W.: Cisco visual networking index: global mobile data traffic forecast update, 2015–2020 white paper. Technical Report, Cisco (2015)

Author Index

© Springer International Publishing AG 2018
R. Lee (ed.), *Software Engineering Research, Management and Applications,*
Studies in Computational Intelligence 722, DOI 10.1007/978-3-319-61388-8